Analog Filters

Kendall L. Su

Regents' Professor Emeritus, School of Electrical and Computer Engineering,
Georgia Institute of Technology, Atlanta, USA.

KLUWER ACADEMIC PUBLISHERS
BOSTON/DORDRECHT/LONDON

Distributors for North, Central and South America:
Kluwer Academic Publishers
101 Philip Drive
Assinippi Park
Norwell, Massachusetts 02061 USA
Telephone (781) 871-6600
Fax (781) 871-6528
E-Mail <kluwer@wkap.com>

Distributors for all other countries:
Kluwer Academic Publishers Group
Distribution Centre
Post Office Box 322
3300 AH Dordrecht, THE NETHERLANDS
Telephone 31 78 6392 392
Fax 31 78 6546 474
E-Mail services@wkap.nl>

 Electronic Services <http://www.wkap.nl>

A catalog record for this book is available from the British Library

ISBN 0-412-63840-1

First edition 1996

Copyright © 1996 Kendall L. Su
Second printing 1999

Printed on acid-free paper.

Contents

Preface

This book is intended as an intermediate-level introduction to the basic theory of analog filters. It covers three major fundamental types of analog filters - passive, active, and switched-capacitor. The only basic knowledge required to follow the material in this book is some basic circuit theory, signal analysis, Laplace transforms, and mathematics typically required by most engineering curriculums at the sophomore level.

The emphasis of this book is on giving the student some fundamental principles behind the various techniques of analog filter design. It is targeted toward students in communications, signal processing, electronics, controls, etc. It is not meant to be an in-depth or comprehensive treatment of the entire area of filter theory as network theorists and filter designers would like to present such a subject. Rather, it is meant to expose the student to the elegant theory behind the development of analog filters. It also introduces the student to the jargon used and techniques practiced in analog filters.

The design of standard filters is now a fairly routine matter. In fact, one can purchase software at a very modest cost and generate filters without ever knowing anything about the underlying principles. The main purpose of this book is to engender some understanding of the mathematical basis of network synthesis and filter theory. Although the mechanical steps for generating filters are covered, they are not the major focus of this book. It stresses the mathematical bases and the scholastic ingenuity of analog filter theory. In other words, the student will learn *why* analog filters work as well as *how* they can be generated.

The book should help nonspecialist electrical engineers in gaining a background perspective and some basic insight into the development of real-time filters. In many modern advances in signal handling, their concepts and procedures have close links to analog filters, either conceptually or mathematically. The material in this book will provide engineers with a better perspective and more penetrating appreciation of these modern techniques.

This book is designed to be used in a three-hour one-semester course. With the proper choice of topics, parts of this book can be used in a three-hour one-quarter course. The proper timing of a course using this book can vary widely depending on the individual student as well as the curriculum. At one extreme, the course can be taken as early as immediately after the first circuits course. At the other, it can serve as a good complement to other courses in a graduate program. At Georgia Tech, this course is nominally offered as a senior elective. However, some juniors and some graduate students also enroll in it. Students who have cursory exposure to filters from courses such as signals and systems or op amp circuits will find this book particularly satisfying in extending their knowledge in filters.

The software MATLAB® (MATLAB is a registered trademark of The MathWorks, Inc., 24 Prime Park Way, Natick, MA 01760-1500, U.S.A. Telephone: (508) 653-1415. Email: info@mathworks.com. Fax: (508) 653-2997. World Wide Web address: http://www.mathworks.com) has been chosen as the standard computational tool for this book. Many examples are first presented in the usual form. Then these examples are also worked using MATLAB. By following these two versions of solutions, students can actually learn MATLAB as they go through the text. However, MATLAB is not absolutely mandatory to working problems in this volume. If MATLAB is not available to the students, the MATLAB steps can be either bypassed or substituted with other software available at the facility.

The preliminary drafts of this book were used at Georgia Tech several times in the form of class notes. The author has been greatly encouraged by his students in these classes in their zeal to acquire knowledge and enthusiasm in taking part in the birth of a new book. They have offered many suggestions and pointed out errors in the preliminary drafts. They have been important contributors to the formation of this book.

Kendall L. Su

Chapter 1

Introduction

This chapter describes the roles and basic concepts of analog filters, their applications in electrical engineering, and certain fundamental ideas and terminology associated with them.

The term *filter* is used in many different ways in electrical engineering. An algorithm in a computer program that makes a decision on which commands and how certain commands are executed performs a filtering function. A decision technique that estimates the input signal from a set of signals and noise is known as optimal filtering. In analog and digital signal processing, filters eliminate or greatly attenuate the unwanted portion of an input signal. These analog and digital filtering processes may be performed in real-time or in off-line situations.

1.1 Preliminary remarks

In this volume, we are dealing with filters that are physical continuous-time subsystems that perform certain input-output relationships in real time. Filters in this context are so widely used in electronic systems, such as electronics, military ordnance, telecommunications, radar, consumer electronics, that it is difficult to imagine any of these applications not containing components that can be identified as filters.

The development of filters started in the early part of the twentieth century. The early progress of filters was primarily associated with applications in telephony. The methods used were somewhat heuristic and empirical. Although later theoreticians could demonstrate that those earlier designs were suboptimal, those earlier achievements were not too far from the theoretical optimum in performance. In other words, when theoreticians were able to accomplish the optimum filter design, the improvements were really quite modest. This really makes the achievements of the earlier filter design without sophisticated mathematical means quite remarkable. However, in most cases, the more modern design techniques, in addition to rendering filters that are optimum in some sense, also employed a synthesis philosophy that is mathematically rigorous and scholastically satisfying.

The major progress in filter theory was largely accomplished in the 1930's and 1940's. The theoretical thoroughness and elegance of this body of knowledge is, to this day, one of the most beautiful and admirable intellectual achievements in electrical engineering.

Those elegant and successful studies of filters were, however, limited to filters made of lossless elements - inductors, capacitors, and mutual inductances. In the late 1940's a new type of analog filters emerged - the active filters. Initial efforts were motivated by the low-frequency applications of filters in which inductors become too costly and their weights and volumes become excessive. In some of those applications, even the use of vacuum tubes could prove to be superior to the use of inductors. Because of the availability of active elements and the advent of solid-state devices and technologies, active filters eventually proved to be practical and attractive under many circumstances. In many situations, active and passive filters are equally adept in their suitability. Under other circumstances, they complement each other. On still other occasions, one type is clearly superior to the other. Hence, both passive and active filters have their places in electronic technology.

The development of active filters has been quite different from that of passive filters. Since passive filters are limited in terms of the types of components used, the available circuits and circuit configurations are rather limited. On the other hand, since there are large numbers of available active devices and configurations, the types of circuits suitable for use as active filters are also very large. In the 1960's and 1970's, literally hundreds of active filter circuits were proposed. For various reasons, mostly practical ones, the commonly used active filters appear to be limited to a few more popular configurations. This is not to say that other circuits cannot perform the same filtering tasks. Rather, these popular circuits are quite adequate for almost all engineering needs. Therefore, there is no reason to try to use other known circuits or invent new ones. Our study of active filters here will be limited to circuits that have been proven practical, are widely used for various reasons, or pedagogically beneficial.

1.2   The analog filter

The term *analog filter* shall be used to mean that branch of filter theory that makes use of linear time-invarying elements to perform certain tasks on continuous-time analog signals. The term analog signals refers to signals that have not been quantized or digitized in their strength as functions of time.

An analog filter is typically a single-input single-output system as shown in Fig. 1.1. In Fig. 1.1(a), the input $x(t)$ and output $y(t)$ are both spec-

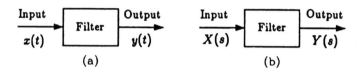

(a)   (b)

Figure 1.1: Representation of a filter (a) in the time domain and (b) in the frequency domain.

ified in the time domain. Both $x(t)$ and $y(t)$ may be either a voltage or a current. For this type of situation, the filter is often referred to as the *pulse-forming network* as it focuses on the waveshape aspects of the output-input relationship. In Fig. 1.1(b), the input-output relationship is governed by the network function $H(s)$ in the complex frequency domain where [1]

$$H(s) = \frac{Y(s)}{X(s)} \tag{1.1}$$

in which $s = \sigma + j\omega$ and $X(s)$ and $Y(s)$ may be regarded as the Laplace transform of $x(t)$ and $y(t)$ respectively.

In this volume, we shall concentrate on the situation formulated in Fig. 1.1(b). In reality, most time-domain application requirements of Fig. 1.1(a) are translated into their equivalent in the frequency domain and then solved as if it were a frequency-domain problem [Su2].

In the frequency domain, our focus is generally directed toward either the *magnitude* and/or the *phase* of the network function on the j axis of the s plane. Or

---

[1] In some filter literature, the reciprocal of this formalism is used. The reader should not have any difficulty in making the distinction between these two conventions.

$$H(s)\Big|_{s=j\omega} = H(j\omega) = |H(j\omega)| \, e^{j\phi(\omega)} \tag{1.2}$$

where $|H(jw)|$ is the *magnitude function* and $\phi(\omega)$ is the *phase function*.

With the formulation of the function in (1.1), the magnitude function is the *gain* function.[2] This function is frequently expressed in dB, viz.

$$\alpha(\omega) = 20\log|H(j\omega)| \text{ dB} \tag{1.3}$$

The phase function given in (1.2) is the phase *lead* function - the phase angle by which the output signal leads the input signal. Another important function is the *group delay*

$$T_d(\omega) = -\frac{d}{d\omega}\phi(\omega) \tag{1.4}$$

The phase function and the group delay function have profound time-domain ramifications as they have a direct effect on the waveshape of the output signals.

As a practical matter, the magnitude functions and the phase functions are usually dealt with separately. The reason for this approach is that the realization of a network function to furnish both a desirable magnitude function and a phase or delay function is simply too difficult, sometimes impossible, mathematically. There is a certain basic and inherent interrelationship between these two functions and they cannot be specified entirely independently.

In some applications, the magnitude function is the only one that matters. In that case, we simply accept the delay function that accompanies the realized magnitude function. If both the magnitude and delay functions are important, we usually prefer to realize the given magnitude function, as best we can, first. Then if the accompanying delay function is not satisfactory, we introduce additional networks, known as *delay equalizers* or *phase linearizers*, to improve the delay function.

1.3 Ideal and approximate filter characteristics

In Fig. 1.2, five ideal filter magnitude characteristics are shown. They are (a) *lowpass*, (b) *highpass*, (c) *bandpass*, (d) *bandreject*, and (e) *allpass* filters. The first four types of characteristics are used for their

---

[2]If $H(s)$ is defined as $X(s)/Y(s)$, then $|H(j\omega)|$ would the *loss function*.

frequency-selective properties. The allpass filter is used chiefly for its phase-linearization or delay-equalization capabilities. These ideal char-

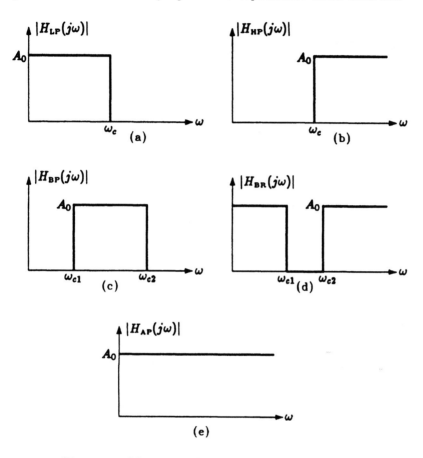

Figure 1.2: Magnitude characteristics of ideal filters.

acteristics are not realizable with finite networks. Hence all real-world filters can have only magnitudes that approximate these characteristics.

Take the approximate lowpass characteristic. We divide the frequency axis into three segments as shown in Fig. 1.3. The region $0 < \omega < \omega_p$ is considered to be the *pass band* in that the gain is relatively high so signals in this range are 'passed through.' Another phenomenon that we are forced to accept is that the gain in this region cannot be constant. Some variation is inevitable. As long as a characteristic is confined to the shaded box bound by $0 < \omega < \omega_p$ and $A_1 < |H(j\omega)| < A_0$, we normally consider it to have the same degree of approximation in the pass band. Two hypothetical characteristics are shown in Fig. 1.3. The

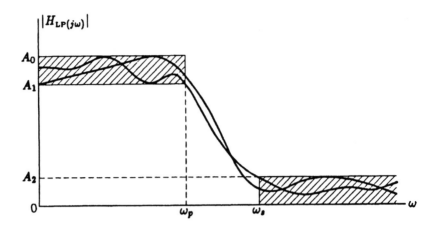

Figure 1.3: Approximate lowpass magnitude characteristics.

amount of variation in the pass band is indicated by the maximum gain variation known as the *passband ripple* or *passband variation*, α_p, where

$$\alpha_p = 20 \log \left[\frac{A_0}{A_1} \right] \text{ dB} \qquad (1.5)$$

At the other end of the ω axis, the range $\omega_s < \omega < \infty$, in which the gain is made very low, is called the *stop band*. Signals in this range are suppressed by the filter as much as practical - ideally with zero gain. Again, some finite gain and variation are inevitable. It is desirable that the maximum gain in this range, A_2, is made as low as practical. All magnitude characteristics that fall within the rectangle $\omega_s < \omega < \infty$ and $0 < |H(j\omega)| < A_2$ are considered to have the same degree of approximation in the stop band. The degree signals in this range are suppressed is indicated by the *stopband attenuation*, α_s, where

$$\alpha_s = 20 \log \left[\frac{A_0}{A_2} \right] \text{ dB} \qquad (1.6)$$

The region $\omega_p < \omega < \omega_s$ is known as the *transition band*. Usually how a characteristic varies in this region is not given too much emphasis in selecting a filter characteristic. The quantity ω_s/ω_p is known as the *transition-band ratio*.

This general convention and terminology are easily extended to highpass, bandpass, and bandreject characteristics. However, in this volume, our primary attention will be paid to the lowpass filters, as there are techniques by which these other three types of filters can be derived from basic lowpass ones.

In general, it is desirable for a filter to have the following characteristics:

1. Small passband ripple or variation, α_p.

2. Large stopband attenuation, α_s.

3. Low transition-band ratio, or $\omega_s/\omega_p \Rightarrow 1$.

4. Simple filter network. This is directly related to the order of the network function $H(s)$.

As with many engineering designs, these goals are mutually conflicting. For example, if we hold the order of the network function fixed, a small α_p can only be achieved at the expense of either a larger transition-band ratio or a lower stopband attenuation. Hence the choice of a suitable filter magnitude characteristic is a matter of compromise or trade-off of these mutually competing properties.

1.4 MATLAB

A brief introduction to the software MATLAB is given here. The version suitable for computational work in this volume is the Student Edition Version 4, which includes some symbolic mathematics tools.[3] For readers who do not have a working knowledge of MATLAB, the user's guide is necessary[?]. We shall give a few examples to show how MATLAB is formulated.

The basic units in MATLAB are matrices. The elements of the matrices may be complex or symbolic. The matrices may be column vectors, row vectors, or scalars. For our purposes, almost all units are vectors or scalars.

For example, if we wish to multiply two polynomials $f_1(s) = 5s^3 + 4s^2 + 2s + 1$ and $f_2(s) = 3s^2 + 5$, MATLAB uses the convolution of two number sequences to obtain the coefficients of the product polynomial. The steps are

```
» f1 = [5 4 2 1];
» f2 = [3 0 5];
» f3 = conv(f1,f2)
f3 = 15 12 31 23 10 5
```

---

[3]Alternatively, one can use the professional version of MATLAB with Signal Processing, Control System, and Symbolic Math Toolboxes.

The final answer f3 is a row vector and renders the result $f_3(s) = f_1(s) \times f_2(s) = 15s^5 + 12s^4 + 31s^3 + 23s^2 + 10s + 5$.

Using the commands from the Symbolic Math Toolbox, we will have

```
≫ f1 = '5*s^3 + 4*s^2 + 2*s + 1';
≫ f2 = '3*s^2 + 5';
≫ f3 = symmul(f1,f2);
≫ f3 = expand(f3)
f3 = 15*s^5 + 12*s^4 + 31*s^3 + 23*s^2 + 10*s + 5
```

Another command that is used frequently in this volume is roots which gives the zeros of a polynomial. Using the row vector of f3 above

```
≫ z = roots(f3)
f3 = 1.2910i, -1.2910i, -0.6553, -0.0723 + 0.5477i,
-0.0723 - 0.5477i
```

It's implicit that the first value of z is z(1), the second value z(2), etc. Conversely, the command poly gives the polynomial coefficients whose zeros are known. Thus

```
≫ f4 = poly([z(3) z(4) z(5)])
f = 1 0.8 0.4 0.2
```

The two vectors f4 and f1 differ only by a constant multiplier. Obviously, they have the same zeros.

A row vector that consists of a number of equally spaced elements can be generated by the command x = a:b:c, in which a is the starting value, b is the spacing between element values, and c is the last value. Hence x = 0:0.5:4 gives x = 0 0.5 1 1.5 2 2.5 3 3.5 4. The default value for b is 1. This command is particularly convenient when a large number of points are to be generated for the plot of a curve.

The reader should refer to the user's guide for other features and basic commands of MATLAB. Those commands that are particularly useful for our purposes will be introduced and explained as we use them throughout the volume.

MATLAB executes each command as it is entered from the keyboard. Those followed by a semicolon are executed but the answers are not displayed. This makes it difficult to debug a program if the program contains a sizable number of statements. Also, when a function is relatively long, it may be difficult to enter it through the keyboard without making mistakes. In those instances, it is best to prepare the sequence of statements with a word processor to create an ASCII file and store it as an 'M-file' - a file with the letter 'm' as its extension. These files can be invoked and executed as if the statements were entered through the keyboard.

The `diary` command initiates a record as a file of all that appears on the screen until it is toggled. The saved file can then be retrieved, edited, and printed. The printing of a file is usually hardware-dependent and the readers should familiarize themselves with the necessary protocol for printing a document.

1.5 Circuit analysis

It is assumed that the reader has a basic knowledge and ability in circuit analysis and is capable of obtaining a network function for a given network. The analysis of a filter circuit is usually the first step in the study of a filter. In dealing with filters, certain analysis techniques are particularly suitable and efficient. We shall refresh the reader's memory with a couple of examples.

EXAMPLE 1.1. Obtain the transfer function $H(s) = E_2/E_1$ of the ladder network shown in Fig. 1.4.

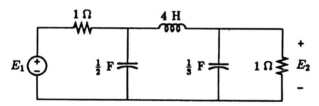

Figure 1.4: Circuit used in Example 1.1.

SOLUTION **1. Node analysis**

We set up the circuit for node analysis as shown in Fig. 1.5.

The two node equations are

$$\frac{E_3 - E_1}{1} + \frac{1}{2}sE_3 + \frac{E_3 - E_2}{4s} = 0$$

$$\frac{E_2 - E_3}{4s} + \frac{1}{3}sE_2 + \frac{E_2}{1} = 0$$

Eliminating E_3 and rearranging, we get

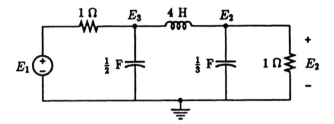

Figure 1.5: Node analysis of the circuit.

$$H(s) = \frac{E_2}{E_1} = \frac{6}{4s^3 + 20s^2 + 29s + 12}$$

This last step can be accomplished by using the `solve` command of MATLAB. The format of this command is

$$\texttt{solve(S1, S2, ... , 'v1, v2, ...')}$$

where $\texttt{Sj} = 0$ are the equations in symbolic form and $\texttt{vi}$ are the variables to be solved. For this example, the MATLAB steps are

```
» f1 = '(E3-E1) + 1/2*s*E3 + (E3-E2)/(4*s)';
» f2 = '(E2-E3)/(4*s) + 1/3*s*E2 + E2';
» solve(f1,f2,'E2,E3')
E2 = 6*E1/(4*s^3 + 20*s^2 + 29*s + 12),
E3 = 2*(3 + 4*s^2 + 12*s)*E1/(4*s^3 + 20*s^2 + 29*s + 12)
```

In this example, we are interested in solving for E_2 in terms of E_1. We are not particularly interested in E_3. In fact, we want to eliminate E_3. Since we have only two equations, we can only solve for two unknowns in terms of all other variables. By solving for E_2 and E_3, we can obtain an expression for E_2 without E_3 in it. Thus E_3 is eliminated from the equation for E_2.

2. Ladder analysis

The ladder network is an important configuration in filters and the analysis of such a network can be performed in a step-by-step manner. We shall reanalyze the circuit of Fig. 1.4 by this procedure. In the arrangement of Fig. 1.6, we have successively

$$I_3 = \frac{E_2}{1} + \frac{s}{3}E_2$$

$$E_4 = E_2 + 4sI_3 = \left(\frac{4}{3}s^2 + 4s + 1\right)E_2$$

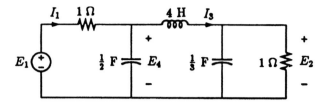

Figure 1.6: Ladder analysis of the circuit.

$$I_1 = I_3 + \frac{1}{2}sE_4 = \left(\frac{2}{3}s^3 + 2s^2 + \frac{5}{6}s + 1\right)E_2$$

$$E_1 = I_1 \times 1 + E_4 = \left(\frac{2}{3}s^3 + \frac{10}{3}s^2 + \frac{29}{6}s + 2\right)E_2$$

Hence

$$H(s) = \frac{E_2}{E_1} = \frac{1}{\frac{2}{3}s^3 + \frac{10}{3}s^2 + \frac{29}{6}s + 2}$$

EXAMPLE 1.2. The triangular symbol in Fig. 1.7 represents an ideal voltage amplifier with a voltage gain equal to 2 and zero input current. Obtain the voltage transfer function $H(s) = E_2/E_1$.

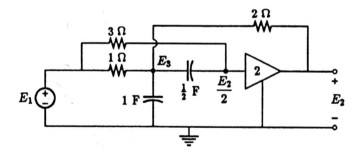

Figure 1.7: Circuit used in Example 1.2.

SOLUTION The various node voltages in the circuit are labeled in the diagram. Two node equations can be written for the circuit. They are

$$\frac{E_3 - E_1}{1} + sE_3 + \frac{s}{2}\left(E_3 - \frac{E_2}{2}\right) + \frac{E_3 - E_2}{2} = 0$$

$$\frac{E_2/2 - E_1}{3} + \frac{s}{2}\left(\frac{E_2}{2} - E_3\right) = 0$$

Eliminating E_3 and rearranging, we get

$$H(s) = \frac{E_2}{E_1} = \frac{4(2s+1)}{2s^2 + 3s + 2}$$

Using MATLAB, we would have the following steps.

```
> f1 = 'E3-E1 + s*E3 + s/2*(E3-E2/2) + (E3-E2)/2';
> f2 = '(E2/2-E1)/3 + s/2*(E2/2-E3)';
> solve(f1,f2,'E2,E3')
E2 = 4*E1*(1 + 2*s)/(2*s^2 + 3*s + 2)
% The answer for E3 has been discarded
```

1.6 Normalization and denormalization - scaling

Since filters are used for many different frequency ranges and impedance levels, it is desirable to be able to standardize the basic theory and design, and then adapt the filter to different applications. This section deals with these considerations.

1.6.1 Frequency scaling

If every inductance and every capacitance in a network is divided by the _frequency scaling factor_, k_f, then the network function $H(s)$ becomes $H(s/k_f)$. What occurs at ω in the original network will now occur at $k_f\omega$.

EXAMPLE 1.3. The network in Fig. 1.8 is a lowpass filter. Analysis will yield the voltage transfer ratio

$$H(s) = \frac{E_2}{E_1} = \frac{0.5}{2.036s^3 + 2.012s^2 + 2.521s + 1}$$

To obtain a plot of its magnitude characteristic we can use the MATLAB commands **bode** and **plot**.

```
% Specify the numerator coefficient vector
> num =0.5;
```

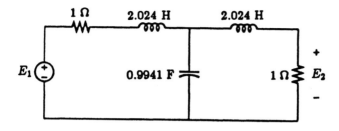

Figure 1.8: A lowpass filter.

```
% Specify the denominator coefficient vector
» den = [2.036 2.012 2.521 1];
% Assign values of ω from 0 to 5 in steps of 0.01
» w=0:0.01:5;
% Compute |H(jω)| (mag) and arg[H(jω)] = φ(ω) (phase) for all
% values of ω
» [mag,phase]=bode(num,den,w);
% obtain a plot of |H(jω)| versus ω
» plot(w,mag)
```

The plot of the frequency characteristic $|H(j\omega)|$ is shown in Fig. 1.9. This filter magnitude characteristic has a value of $\omega_p = 1$ rad/sec. We apply a frequency scaling factor $k_f = 100$. The scaled network is shown in Fig. 1.10. The voltage ratio function of the scaled network is shown in Fig. 1.11.

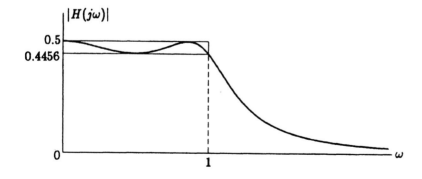

Figure 1.9: Voltage ratio characteristic of the filter in Fig. 1.8.

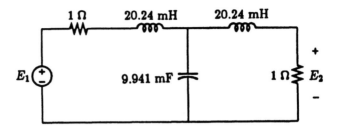

Figure 1.10: The filter in Fig. 1.8 scaled by $k_f = 100$.

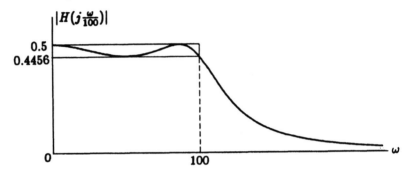

Figure 1.11: Voltage ratio characteristic of the filter in Fig. 1.10.

1.6.2   Impedance scaling

If the value of every element in a network that has the dimension of the resistance (this includes the multiplicative constant of a current-controlled voltage source) or inductance (this includes the self and mutual inductances) is multiplied by the *impedance scaling factor*, k_z; and the value of every element that has the dimension of the conductance (this includes the multiplicative constant of a voltage-controlled current source) or capacitance is divided by k_z, then the following consequences occur.

(1) All network functions that have the dimension of the impedance (such as driving-point impedances $Z_{\text{in}} = E_1/I_1$ and the transfer impedance $Z_{21} = E_2/I_1$) will be multiplied by k_z.

(2) All network functions that have the dimension of the admittance (such as the driving-point admittance $Y_{\text{in}} = I_1/E_1$ and the transfer admittance $Y_{21} = I_2/E_1$) will be divided by k_z.

(3) All network functions that are dimensionless (such as the voltage ratio E_2/E_1 and the current ratio I_2/I_1) will be unaffected.

EXAMPLE 1.4. If we take the filter of Fig. 1.10 and impedance-scale it by $k_z = 4$, the network of Fig. 1.12 results. This new network will have the same voltage transfer characteristic shown in Fig. 1.11.

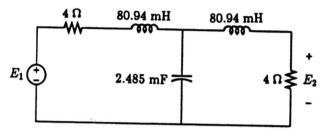

Figure 1.12: The filter in Fig. 1.10 impedance-scaled by $k_z = 4$.

1.6.3 Normalization and denormalization

Since we are able to frequency-scale and impedance-scale a filter, we usually prefer to study and design filters in their normalized form. For example, we may choose to normalize the pass band by making $\omega_p = 1$ rad/sec. We may choose one of the resistances to be 1 ohm or one of the capacitances to be 1 farad. Normalization of networks makes it possible for us to deal with numbers that are much easier to handle. They are generated in the anticipation that they will eventually be frequency-scaled and impedance-scaled to suit the particular application in hand.

1.7   Steps involved in the design of filters

The design of a filter usually starts with a set of parameters that will satisfy the requirements of an application. We shall call this step the *specification*. For example, it may be necessary to obtain a lowpass filter for the situation of Fig. 1.13, in which the transfer function in question is $H(s) = E_2/E_1$. The pass band is $\omega_p = 2\pi \times 10^3$ rad/sec (1 kHz). The stop band starts at $\omega_s = 2\pi \times 1.5 \times 10^3$ rad/sec (1.5 kHz). The passband ripple is $\alpha_p = 2$ dB. The stopband attenuation is $\alpha_s \geq 25$ dB.

We may prefer to design an alternative filter that is normalized. We can normalize the load resistance to 1 Ω as shown in Fig. 1.14. We may further normalize the magnitude and frequency scales such that $\omega_p = 1$ rad/sec, $\omega_s = 1.5$ rad/sec, and $H(0) = 1$.

The next step is to find an $|H(j\omega)|$ that satisfies these requirements. For example, the following magnitude-squared function will satisfy these

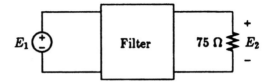

Figure 1.13: The arrangement of a filter application.

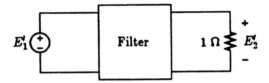

Figure 1.14: Filter of Fig. 1.13 with the load normalized.

normalized requirements:

$$|H(j\omega)|^2 = 1.5849/(1.5849 - 9.3583\omega^2$$

$$+46.7914\omega^4 - 74.8663\omega^6 + 37.4331\omega^8) \tag{1.7}$$

The variation of $|H(j\omega)|$ in (1.7) is shown in Fig. 1.15. This step is known as *approximation*.

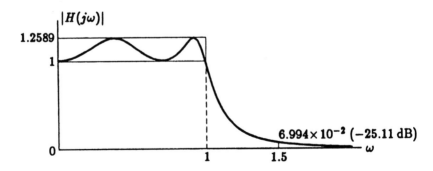

Figure 1.15: Variation of $|H(j\omega)|$ in (1.7).

The next step is to obtain a network function whose magnitude is equal to the approximating characteristic. In the example above the network function is

$$H(s) = \frac{1}{4.8598s^4 + 3.4807s^3 + 6.1062s^2 + 2.5115s + 1} \qquad (1.8)$$

After the network function has been determined, the next task is to find a suitable network that will realize this network function. This step draws heavily on the theory and techniques developed in the area of *network synthesis*, both active and passive. Usually the solution of this step is not unique in that many networks can be used to realize the same network function. For the examples stated above the circuits of Fig. 1.16 and Fig. 1.17, as well as many others, will realize the required network function given in (1.8). This step is known as *realization* or *synthesis*.

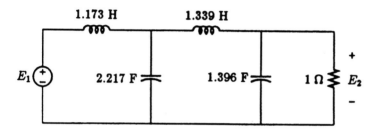

Figure 1.16: A passive circuit that realizes the $H(s)$ in (1.8).

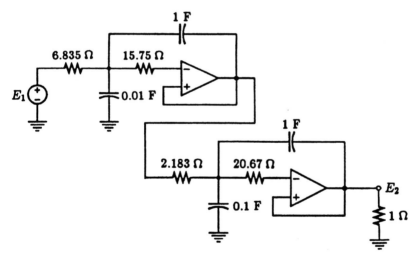

Figure 1.17: An active circuit that realizes the $H(s)$ in (1.8).

The last step is to apply the appropriate frequency scaling factor, k_f, and the appropriate impedance scaling factor, k_z, to suit the application needs. In the example above, if the original requirements are to be met, we would apply $k_f = 2\pi \times 10^3$ and $k_z = 75$ to the normalized circuits in Fig. 1.16. For the circuit in Fig. 1.17 we would also apply $k_f = 2\pi \times 10^3$. Since the output is also the output of the second op amp, the voltage ratio E_2/E_1 is independent of the load impedance. We can simply change the value of the load resistance from 1 ohm to 75 ohms. We can apply any k_z to the op amp circuit to make the element values more practical. This choice of k_z will not affect the voltage ratio. This last step of frequency and impedance scalings is known as *denormalization*.

The sequence of steps is summarized in the block diagram of Fig. 1.18.

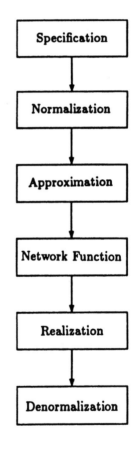

Figure 1.18: Usual steps involved in the design of filters.

Problems

1.1 Derive the transfer function, $H(s) = E_2/E_1$ or E_2/I_1, for the following networks.

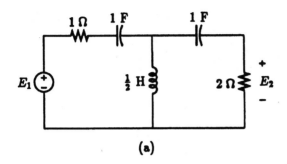

(a)

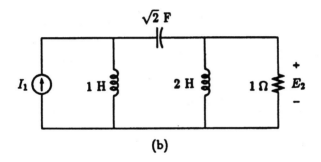

(b)

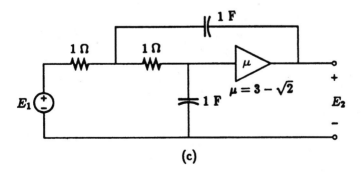

(c)

1.2 Use MATLAB to derive the transfer function $H(s) = E_2/E_1$ of the following circuit.

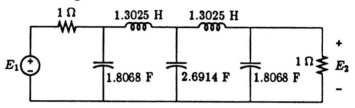

1.3 Use MATLAB to derive the transfer function $H(s) = E_2/E_1$ of the following circuit.

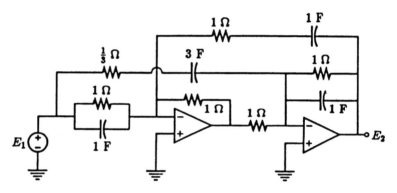

1.4 Use MATLAB to produce plots of $|H(j\omega)|$ and $\phi(\omega)$ from $\omega = 0$ to $\omega = 4$ for the following network functions.

(a) $H(s) = \dfrac{0.4913}{s^3 + 0.9883s^2 + 1.2384s + 0.4913}$

(b) $H(s) = \dfrac{0.1614(s^2 + 1.1536)(s^2 + 3.3125)}{(s^2 + 0.07393s + 1.0107)(s^2 + 0.8019s + 0.6849)}$

1.5 (a) Derive the voltage transfer function, $H(s) = E_2/E_1$, for the bandpass filter shown below. (b) Give an accurate plot of $|H(j\omega)|$ for $0 < \omega < 2$.

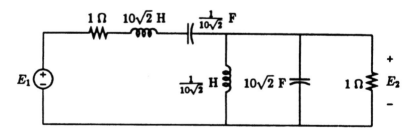

1.6 The following circuit is a lowpass filter with $\alpha_p \approx 0.25$ dB and $\omega_p = 1$ rad/sec. Modify the circuit so that it becomes a lowpass filter with a pass band of 10 kHz and a load resistance of 75 Ω.

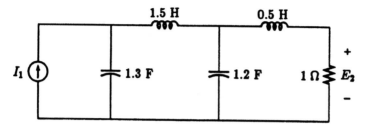

1.7 It is known that the following network realizes the voltage ratio

$$H(s) = \frac{E_2}{E_1} = \frac{2728.4s}{s^2 + 100s + 10^6}$$

which is a bandpass filter with the band center located at 1000 rad/sec. First, apply a frequency scaling to normalize the band center to 1 rad/sec. Then apply an impedance scaling to normalize the network such that both capacitors become 1 F.

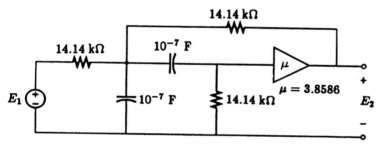

1.8 The following function provides a simple hump or dip in the magnitude function and is sometimes used to equalize the gain in the neighborhood of a certain frequency. Give a plot of $|H(j\omega)|$ for (a) $a = 0.24$ and (b) $a = 0.16$.

$$H(s) = \frac{s^2 + as + 1}{s^2 + 0.2s + 1}$$

1.9 Obtain the phase and the group delay functions of the following transfer function. Give a plot of the group delay function in the range $0 < \omega < 5$.

$$H(s) = \frac{s^2 - 2s + 4}{s^2 + 2s + 4}$$

Chapter 2

The approximation

As was mentioned in Chapter 1, the first step in the design of a normalized filter is to find a magnitude characteristic, $|H(j\omega)|$, such that the set of specifications of an application is satisfied. Usually it is more convenient to deal with $|H(j\omega)|^2$ instead. As should be well known to the reader, a network function, $H(s)$, must be a real rational function in s (the ratio of two polynomials with real coefficients), $|H(j\omega)|^2$ can be obtained by

$$|H(j\omega)|^2 = H(s)H(-s)\Big|_{s=j\omega} = H(j\omega)H(-j\omega) = H(j\omega)H^*(j\omega) \quad (2.1)$$

Hence $|H(j\omega)|^2$ is an even function and is a rational function in ω^2. For example, if

$$H(s) = \frac{s+2}{s^3 + 2s^2 + 2s + 3}$$

$$H(s)H(-s) = \frac{s^2 - 4}{s^6 - 8s^2 - 9}$$

$$|H(j\omega)|^2 = \frac{\omega^2 + 4}{\omega^6 - 8\omega^2 + 9}$$

Mathematically it's a great deal easier to deal with $|H(j\omega)|^2$ instead of the irrational function

$$|H(j\omega)| = \frac{\sqrt{\omega^2 + 4}}{\sqrt{\omega^6 - 8\omega^2 + 9}}$$

The study of methods of finding a real rational function in ω^2 to approximate an arbitrary magnitude or magnitude-squared characteristic is a rather specialized area. Numerous techniques using various criteria to achieve the approximation are available. These general techniques are usually needed when the magnitude characteristics do not have any standard pattern. The approximation of these general magnitude characteristics is more for magnitude equalization (as opposed to filtering) purposes. For our purposes, we are mostly interested in the magnitude characteristics that approximate the ideal filter characteristics of various types shown in Fig. 1.2. For convenience, we shall concentrate on the characteristics that approximate the ideal lowpass characteristic. In particular, we shall deal with those types of nonideal lowpass characteristics that are in common use - the Butterworth, the Chebyshev, and the elliptic-function magnitude characteristics.

The general approach in obtaining a lowpass characteristic is to seek a function of the form

$$|H(j\omega)|^2 = \frac{A_0}{1 + F(\omega^2)} \tag{2.2}$$

such that

$$F(\omega^2) \ll 1 \qquad 0 < \omega < \omega_p$$

$$F(\omega^2) \gg 1 \qquad \omega > \omega_s$$

This, in turn, makes $|H(j\omega)|^2 \approx A_0$ and $|H(j\omega)|^2 \leq A_0$ in the pass band and $|H(j\omega)|^2 \ll A_0$ in the stop band. These are the essential features of a lowpass characteristic that approximates the ideal lowpass characteristic.

2.1   The Butterworth lowpass characteristic

One of the simplest lowpass magnitude characteristics was first suggested by Butterworth in 1930 [Bu]. Because of its simplicity, it is often used when the filtering requirement is not too demanding.

2.1.1   The normalized Butterworth lowpass characteristic

One of the simplest choices for $F(\omega^2)$ in (2.2) is to make

$$F(\omega^2) = \omega^{2n} \tag{2.3}$$

where n is a positive integer. The magnitude-squared characteristic is

$$|H(j\omega)|^2 = \frac{1}{1 + \omega^{2n}} \tag{2.4}$$

A filter that satisfies (2.4) is known as the *Butterworth filter of the nth order*. It is clear that

$$|H(j\omega)|_{\max} = H(0) = 1 \tag{2.5}$$

for any n. As ω increases, $|H(j\omega)|$ decreases monotonically. Also, for any n,

$$|H(j1)|^2 = \frac{1}{2} \tag{2.6}$$

Hence any $|H(j\omega)|$ given in (2.4) has the value $1/\sqrt{2}$ at $\omega = 1$. The gain at this point is 3.0103 dB below the maximum gain. This point is commonly referred to as the 3-dB or half-power point.

If we apply the binomial expansion of $|H(j\omega)|$, we can write

$$|H(j\omega)| = (1 + \omega^{2n})^{-\frac{1}{2}} = 1 - \frac{1}{2}\omega^{2n} + \frac{3}{8}\omega^{4n} - \frac{5}{16}\omega^{6n} + \cdots \tag{2.7}$$

in the vicinity of $\omega = 0$. The first $2n-1$ derivatives of $|H(j\omega)|$ are zero at $\omega = 0$. Since $F(\omega^2)$ is of degree $2n$ in ω and we have made $|H(j0)| = 1$, (2.7) shows that we have made the $|H(j\omega)|$ curve as flat as possible at $\omega = 0$. This characteristic is often referred to as the *maximally flat magnitude characteristic*. Hence, in the range $0 < \omega < 1$, the higher n is, the flatter the characteristic is at the origin, and it approaches the ideal lowpass characteristic of Fig. 1.2(a) more closely. For $\omega > 1$, the higher n is, the faster ω^{2n} increases and the faster $|H(j\omega)|$ decreases as ω is increased. Fig. 2.1 shows the $|H(j\omega)|$ characteristics for several values of n.

For $\omega \gg 1$, the Butterworth magnitude characteristic may be approximated by

$$|H(j\omega)|^2 \approx \frac{1}{\omega^{2n}} \tag{2.8}$$

$$\alpha \approx -10\log(\omega^{2n}) = -20n\log\omega \tag{2.9}$$

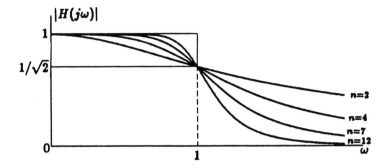

Figure 2.1: Normalized Butterworth magnitude characteristics for several values of n.

Hence the gain decreases at the rate of $20n$ dB/decade. The magnitude characteristics of Fig. 2.1 plotted as dB versus $\log \omega$ - the Bode plots - are shown in Fig. 2.2.

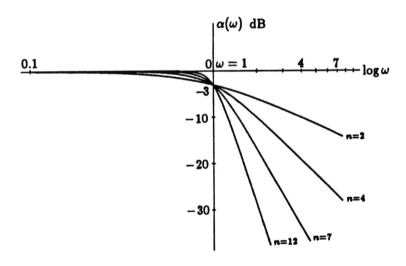

Figure 2.2: The Bode plots of the Butterworth magnitude characteristics of Fig. 2.1.

2.1.2 Using a normalized Butterworth characteristic for a filtering requirement

The squared magnitude function defined in (2.4) has the property that $|H(j1)| = 1/\sqrt{2}$ for any n. If the required filter has a passband variation of 3 dB, then we can simply make $\omega_p = 1$ rad/sec. In practice, the passband variation is not always 3 dB. Hence we need to place ω_p and ω_s at the proper points on the frequency axis as shown in Fig. 2.3. The positioning of ω_p and ω_s is affected by the value of n. Hence we need

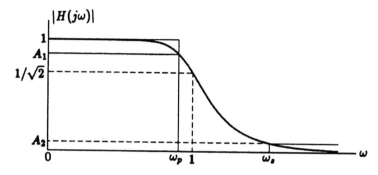

Figure 2.3: Locations of ω_p and ω_s for a Butterworth lowpass characteristic.

to adjust ω_p, ω_s, and n simultaneously. This necessary manipulation is best illustrated by an example.

EXAMPLE 2.1. Determine the Butterworth lowpass characteristic with the minimum n such that the following specifications are satisfied.

$$\alpha_p = 1 \text{ dB} \qquad \alpha_s = 25 \text{ dB} \qquad \omega_s/\omega_p = 1.5$$

SOLUTION   We need

$$10\log(1 + \omega_p^{2n}) = 1 \tag{2.10}$$

$$10\log(1 + \omega_s^{2n}) = 25 \tag{2.11}$$

which leads to

$$\omega_p^{2n} = 10^{0.1} - 1 = 0.25893$$

$$\omega_s^{2n} = 10^{2.5} - 1 = 315.228$$

Since $\omega_s/\omega_p = 1.5$, it is necessary that

$$\left[\frac{\omega_s}{\omega_p}\right]^{2n} = 1.5^{2n} = \frac{315.228}{0.25893} = 1217.4$$

Solving, we get

$$n = 8.761$$

Since n must be an integer, we let

$$n = 9$$

From (2.10), we find[1]

$$\omega_p = 0.9277$$

For this value of ω_p and from (2.4) we get

$$\alpha_s = 10\log[1 + (1.5 \times 0.9277)^{18}] = 25.84 \text{ dB}$$

If we wish to relocate ω_p to 1 rad/sec we simply apply a frequency scaling factor $k_f = 1/0.9277$. The adjusted Butterworth lowpass magnitude characteristic function would be

$$|H_n(j\omega)|^2 = \frac{1}{1 + (0.9277\omega)^{18}}$$

2.2   The Chebyshev lowpass characteristic

Another commonly used standard lowpass characteristic is the Chebyshev lowpass characteristic. This class of characteristics makes use of the Chebyshev polynomial and produces an equal-ripple variation in the pass band. Outside the pass band, the gain also decreases monotonically, but at a faster rate than the Butterworth characteristics.

---

[1] With this choice, α_p is satisfied exactly and α_s is greater than specified. Some latitude is available here since n has been rounded off to the next higher integer.

2.2.1 The Chebyshev polynomial

The Chebyshev polynomial (of the first kind) of the nth order is defined as

$$C_n(\omega) = \cos(n \cos^{-1} \omega) \tag{2.12}$$

This expression has the appearance of a trigonometric function. Actually, it is a polynomial in ω of the nth degree. This can best be seen by looking at the recursive formulas for this class of polynomials. First let

$$\cos \theta = \omega \quad \text{and} \quad \theta = \cos^{-1} \omega \tag{2.13}$$

Then

$$C_{n+1}(\omega) = \cos(n\theta + \theta) = \cos n\theta \cos \theta - \sin n\theta \sin \theta \tag{2.14}$$

$$C_{n-1}(\omega) = \cos(n\theta - \theta) = \cos n\theta \cos \theta + \sin n\theta \sin \theta \tag{2.15}$$

Adding (2.14) and (2.15), we get

$$C_{n+1}(\omega) + C_{n-1}(\omega) = 2 \cos n\theta \cos \theta \tag{2.16}$$

But

$$\cos n\theta = C_n(\omega) \quad \text{and} \quad \cos \theta = \omega$$

Hence (2.16) may be written as

$$C_{n+1}(\omega) + C_{n-1}(\omega) = 2\omega C_n(\omega) \tag{2.17}$$

or

$$C_{n+1}(\omega) = 2\omega C_n(\omega) - C_{n-1}(\omega) \tag{2.18}$$

Thus, if $C_n(\omega)$ and $C_{n-1}(\omega)$ are both polynomials, $C_{n+1}(\omega)$ will also be a polynomial. Further, any Chebyshev polynomial may be expressed in terms of the two polynomials of degrees immediately lower than itself. Since we have

$$C_0(\omega) = \cos(0) = 1$$

Table 2.1: **Chebyshev polynomials of order 0 to 10**

n	$C_n(\omega)$
0	1
1	ω
2	$2\omega^2 - 1$
3	$4\omega^3 - 3\omega$
4	$8\omega^4 - 8\omega^2 + 1$
5	$16\omega^5 - 20\omega^3 + 5\omega$
6	$32\omega^6 - 48\omega^4 + 18\omega^2 - 1$
7	$64\omega^7 - 112\omega^5 + 56\omega^3 - 7\omega$
8	$128\omega^8 - 256\omega^6 + 160\omega^4 - 32\omega^2 + 1$
9	$256\omega^9 - 576\omega^7 + 432\omega^5 - 120\omega^3 + 9\omega$
10	$512\omega^{10} - 1280\omega^8 + 1120\omega^6 - 400\omega^4 + 50\omega^2 - 1$

$$C_1(\omega) = \cos(\cos^{-1} \omega) = \omega$$

we can write successively

$$C_2(\omega) = 2\omega C_1(\omega) - C_0(\omega) = 2\omega^2 - 1$$

$$C_3(\omega) = 2\omega C_2(\omega) - C_1(\omega) = 2\omega(2\omega^2 - 1) - \omega = 4\omega^3 - 3\omega$$

etc. Higher-degree Chebyshev polynomials are listed in Table 2.1.

The Chebyshev polynomials are really the trigonometric identities that express $\cos n\theta$ as a function of $\cos \theta = \omega$. Hence they can be derived directly. For instance, to obtain $C_{11}(\omega)$, we can use the following MAT-LAB steps.

```
» C11 = 'cos(11*acos(w))';
» C11 = pretty(expand(C11))
C11 = 1024w^11 - 2816w^9 + 2816w^7 - 1232w^5 + 220w^3 - 11w
```

To see how $C_n(\omega)$ varies in the range $0 < \omega < 1$, we note that

Table 2.2: **Variation of $C_n(\omega)$ in the range $0 \leq \omega \leq 1$**

ω	$\cos^{-1}\omega$	$n\cos^{-1}\omega$	$C_n(\omega)$
1	0	0	1
.	.	$\frac{\pi}{2}$	0
.	.	π	-1
.	.	$\frac{3\pi}{2}$	0
.	.	2π	1
.	.	$\frac{5\pi}{2}$	0
.	.	3π	-1
.	.	.	.
.	.	.	.
0	$\frac{\pi}{2}$	$n\frac{\pi}{2}$	0 or ± 1

$$C_n(1) = \cos(n \cos^{-1} 1) = \cos 0 = 1$$

If we vary ω from 1 to 0, the various intermediate values of $C_n(\omega)$ will vary as shown in Table 2.2. The last value in Table 2.2 depends on the value of n. The variations for $C_4(\omega)$ and $C_7(\omega)$ are shown in Fig. 2.4.

For $\omega > 1$, (2.12) is more conveniently expressed as

$$C_n(\omega) = \cos(n \cos^{-1} \omega) = \cosh(n \cosh^{-1} \omega) \tag{2.19}$$

Since both $\cosh x$ and $\cosh^{-1} x$ are monotonically increasing functions of x, $C_n(\omega)$ is also monotonic.

For $\omega \gg 1$, from (2.12) and Table 2.1, it can be easily observed that

$$C_n(\omega) \approx 2^{n-1}\omega^n \tag{2.20}$$

From (2.12), we see that

$$
\begin{aligned}
C_n(-\omega) &= \cos[n \cos^{-1}(-\omega)] \\[2mm]
&= \cos[n(\pi + \cos^{-1}\omega)] = \cos n\pi \cos(n \cos^{-1}\omega)
\end{aligned}
$$

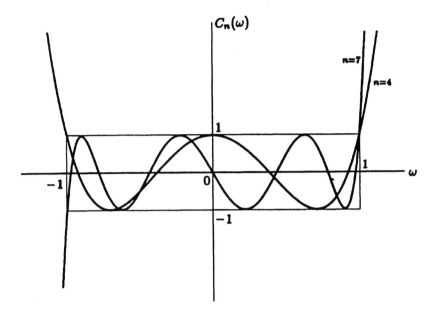

Figure 2.4: Plots of $C_4(\omega)$ and $C_7(\omega)$.

Or

$$C_n(-\omega) = (-1)^n C_n(\omega) \tag{2.21}$$

Thus, $C_n(\omega)$ is an even function if n is even and an odd function if n is odd. This is exemplified in the plots in Fig. 2.4 for $n = 4$ and $n = 7$.

2.2.2   The Chebyshev lowpass characteristic

Since $|C_n(\omega)| \leq 1$ for $|\omega| \leq 1$, we choose a small number ϵ and let

$$F(\omega^2) = \epsilon^2 C_n^2(\omega)$$

then

$$|H(j\omega)|^2 = \frac{1}{1 + \epsilon^2 C_n^2(\omega)} \tag{2.22}$$

will have values that fall between 1 and $1/(1+\epsilon^2)$ in the range $0 \leq \omega \leq 1$. The plots of $|H(j\omega)|$ for $n = 4$ and $n = 7$ are shown in Fig. 2.5. Thus it is quite natural to let $\omega_p = 1$ for the normalized Chebyshev magnitude characteristics and

$$\alpha_p = 10\log(1 + \epsilon^2) \tag{2.23}$$

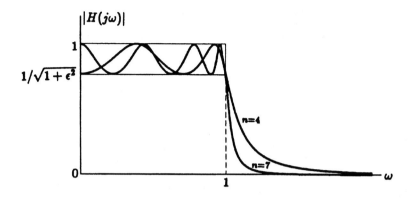

Figure 2.5: Chebyshev characteristics for $n = 4$ and $n = 7$.

For $\omega \gg 1$, from (2.20),

$$|H(j\omega)|^2 \approx \frac{1}{\epsilon^2 2^{2(n-1)}\omega^{2n}} \tag{2.24}$$

Hence the gain is

$$\alpha(\omega) \approx -10\log(\epsilon^2 2^{2(n-1)}\omega^{2n})$$

$$= -20\log\epsilon - 20(n-1)\log 2 - 20n\log\omega \tag{2.25}$$

The first term usually has a positive value as ϵ is usually much less than 1. The second term is nominally $-6(n-1)$ dB. The third term is identical to the right-hand side of (2.9) given for the nth-order Butterworth magnitude characteristic. Thus, for the same passband ripple, the Chebyshev magnitude characteristic attenuates faster than the Butterworth characteristic outside the pass band.

EXAMPLE 2.2. Determine the minimum n of a Chebyshev characteristic necessary to give a magnitude characteristic such that

$$\alpha_p = 1 \text{ dB} \qquad \alpha_s = 25 \text{ dB} \qquad \omega_s/\omega_p = 1.5$$

SOLUTION We have

$$10\log(1 + \epsilon^2) = 1$$

Solving, we get

$$\epsilon^2 = 0.258925$$

Since $\omega_p = 1$, we have $\omega_s = 1.5$ and it is required that

$$10 \log[1 + \epsilon^2 C_n^2(1.5)] = 25$$

Whence

$$C_n^2(1.5) = 1217.45$$

and

$$C_n(1.5) = \cosh(n \, \cosh^{-1} 1.5) = \sqrt{1217.4} = 34.8919$$

$$n = \frac{\cosh^{-1} 34.89}{\cosh^{-1} 1.5} = 4.41094$$

Since n must be an integer, we let $n = 5$. This is considerably lower than the $n = 9$ obtained in Example 1 in which a Butterworth magnitude characteristic is used.

EXAMPLE 2.3. We have a Chebyshev magnitude characteristic with $\alpha_p = 0.5$ dB and $n = 6$. Calculate its attenuation at $\omega = 1.1$, 1.25, and 2.

SOLUTION   We have

$$10 \log(1 + \epsilon^2) = 0.5$$

which requires that

$$\epsilon^2 = 0.12202$$

Hence

$$-\alpha(1.1) = 10 \log[1 + \epsilon^2 C_6^2(1.1)] = 8.64 \text{ dB}$$

$$-\alpha(1.25) = 10 \log[1 + \epsilon^2 C_6^2(1.25)] = 21.00 \text{ dB}$$

$$-\alpha(2) = 10 \log[1 + \epsilon^2 C_6^2(2)] = 53.48 \text{ dB}$$

2.3 Other Chebyshev-related characteristics

If we take the magnitude characteristic defined in (2.22) and define another function

$$|H_a(j\omega)|^2 = 1 - |H(j\omega)|^2 \qquad (2.26)$$

the new magnitude characteristic $|H_a(j\omega)|$ can be used as a highpass characteristic. For $n = 7$, this new characteristic is shown in Fig. 2.6.

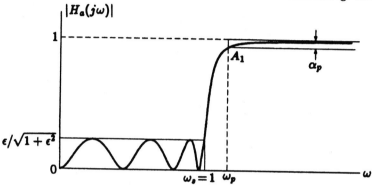

Figure 2.6: Highpass characteristic with equal-ripple variation in the stop band.

The magnitude characteristic $|H_a(j\omega)|$ is the complement of the Chebyshev lowpass characteristic. It has equal-ripple variation in the low-frequency stop band. The stopband attenuation is $10 \log(\epsilon^2/(1 + \epsilon^2))$ and the high-frequency pass band can be placed somewhere in the range $\omega > 1$ as symbolically represented in Fig. 2.6.

If we replace ω in (2.22) by $1/\omega$, we get another magnitude-squared characteristic

$$|H_b(j\omega)|^2 = \frac{1}{1 + \epsilon^2 C_n^2(1/\omega)} \qquad (2.27)$$

This new $|H_b(j\omega)|$ has the same equal-ripple variation with the same α_p in the range $1 \leq \omega < \infty$. This range can now be considered to be the pass band. Hence we have a highpass characteristic with equal-ripple variation in the high-frequency pass band. The stop band should now be placed somewhere in the range $0 < \omega < 1$ as appropriate. For $n = 7$, this characteristic is shown in Fig. 2.7.

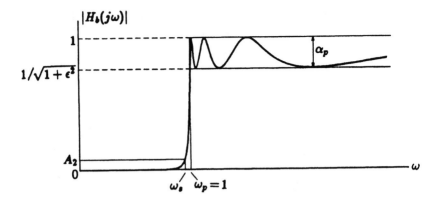

Figure 2.7: Highpass characteristic with equal-ripple variation in the high-frequency pass band.

If we form the complement of the magnitude-squared characteristic of (2.27)

$$|H_c(j\omega)|^2 = 1 - |H_b(j\omega)|^2 \tag{2.28}$$

a lowpass characteristic $|H_c(j\omega)|$ with equal-ripple variation in the stop band is obtained. The stop band is now the range $1 < \omega < \infty$ and the pass band will be located somewhere in the range $0 \le \omega < 1$. An example for $n = 7$ is shown in Fig. 2.8. This characteristic is known as the *inverse Chebyshev characteristic*.

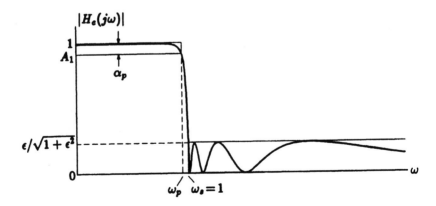

Figure 2.8: The lowpass inverse Chebyshev characteristic.

2.4 The elliptic-function filter characteristic

The numerators of both the Butterworth and the Chebyshev lowpass characteristic are constants. As we shall see in the next chapter, the network functions that have these magnitudes also have constants in their numerators. These functions are known as the *all-pole* functions - all their finite singularities are poles.

We shall now look into another standard lowpass characteristic that has equal-ripple variation in both the pass band and the stop band. For this type of characteristic, we choose the $F(\omega^2)$ of (2.2) to be

$$F(\omega^2) = \epsilon^2 R_N^2(\omega)$$

where

$$R_N(\omega) = \frac{(\omega_1^2 - \omega^2)(\omega_2^2 - \omega^2)\cdots(\omega_N^2 - \omega^2)}{(1 - \omega_1^2\omega^2)(1 - \omega_2^2\omega^2)\cdots(1 - \omega_N^2\omega^2)} \tag{2.29}$$

or

$$R_N(\omega) = \frac{\omega(\omega_1^2 - \omega^2)(\omega_2^2 - \omega^2)\cdots(\omega_N^2 - \omega^2)}{(1 - \omega_1^2\omega^2)(1 - \omega_2^2\omega^2)\cdots(1 - \omega_N^2\omega^2)} \tag{2.30}$$

It is clear that (2.29) is an even function and (2.30) is an odd function. Note that the poles and zeros in (2.29) and (2.30) are chosen such that they are reciprocals of one another. Then we construct the magnitude-squared function

$$|H(j\omega)|^2 = \frac{1}{1 + \epsilon^2 R_N^2(\omega)} \tag{2.31}$$

If we choose $\omega_1, \omega_2, \cdots, \omega_N$ to be in the range $0 \le \omega < 1$, $R_N^2(\omega)$ will have zero values at $\omega_1, \omega_2, \cdots, \omega_N$ and infinite values at $1/\omega_1$, $1/\omega_2, \cdots, 1/\omega_N$. We note that when $R_N(\omega) = 0$, $|H(j\omega)| = 1$; and when $R_N(\omega) = \infty$, $|H(j\omega)| = 0$. Hence the variation of $|H(j\omega)|$ must have the general patterns shown in Fig. 2.9. The plot in Fig. 2.9(a) will result in an $H(s)$ that is of the sixth order, while that in Fig. 2.9(b) will result in one of the seventh order. Thus another type of lowpass characteristics with ripples in both the pass band and the stop band has been created.

The arbitrary placement of ω_i, $i = 1, 2, \cdots, N$, produces the same maxima (equal to unity) in a range within $0 \le \omega < 1$ and same minima

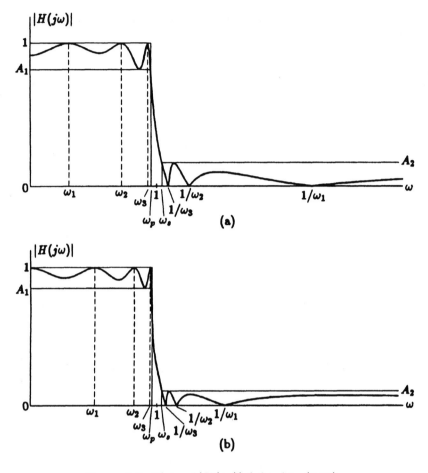

Figure 2.9: Plots of $|H(j\omega)|$ defined in (2.31).

(equal to zero) in a range within $1 < \omega < \infty$. However, the minima in the low-frequency range are not the same. Neither are the maxima in the high-frequency range. It had been shown that if we make these minima and maxima uniform in the respective ranges, we achieve the smallest transition band ratio ω_s/ω_p for the same stopband ripple α_p and passband attenuation α_s. When this is achieved, the $R_N(\omega)$ defined in (2.29) and (2.30) is known as the *Chebyshev rational function*.

It is not a simple matter to determine the locations of ω_i's such that equal-ripple variation will occur in both regions. There are several methods by which this can be accomplished either numerically or by analytical means. We will not attempt to cover these here.

In both the Butterworth and the Chebyshev characteristics, the magnitude functions can be obtained first without paying any attention to what the resulting network function may turn out to be. On the other hand, to obtain equal-ripple variations in both the pass band and the stop band, it is more expedient to obtain the network function $H(s)$ directly first. This will be discussed briefly in Chapter 3.

Filters that have equal-ripple variations in both the pass band and the stop band were first advanced by Cauer in 1931 [Ca]. Filters that have this type of characteristic are called *Cauer filters*. Also, in the derivation of these network functions, the Jacobian elliptic functions can be used. Hence they are also called the *elliptic-function filters*.

2.5 Comparison of standard lowpass characteristics

The Butterworth lowpass magnitude characteristic has the property that it is flattest at $\omega = 0$. Because of this, all available degrees of freedom are utilized to make as many derivatives as possible zero at that point. One consequence of this property is the fact that the rate of increase of attenuation for higher ω is rather low. The chief attribute of this class of characteristics is the simplicity of their mathematical development. However, in accepting this mathematical simplicity, we pay a high price in the relatively poor performance of these filters. This class of characteristics is usually used only when the filtering requirement is very undemanding - high passband variation, low stopband attenuation, or large transition-band ratio.

The Chebyshev lowpass magnitude characteristics exhibit equal-ripple variation in the pass band and monotonic increase in attenuation outside the pass band. The performance of these characteristics is greatly improved over the Butterworth characteristics. It has been shown that if the numerator of $|H(j\omega)|$ is a constant, Chebyshev characteristics give the most rapid increase in attenuation outside the pass band. For the same n, the filter complexity of both the Butterworth and the Chebyshev filters is the same. Hence, the Chebyshev filters are always preferable from the standpoint of the magnitude performance of a network of a given complexity.

The elliptic-function filter characteristic has equal-ripple variation in both the pass band and the stop band. It offers further improvement in the magnitude characteristic for a given order n. In fact, it has been shown that for a given α_p and a given α_s, the elliptic-function characteristic gives the lowest possible transition-band ratio, ω_s/ω_p, of all lowpass magnitude characteristics. The structure of an elliptic-function filter is usually only slightly more complicated than a Butterworth or Chebyshev filter of the same order.

The following numerical examples will demonstrate the relative merits of these three types of magnitude characteristics.

EXAMPLE 2.4. For $n = 7$, $\alpha_p = 1$ dB, and $\alpha_s = 30$ dB, determine the transition-band ratios of the three types of lowpass characteristics.

SOLUTION   (a) For the Butterworth characteristic

$$\omega_p^{14} = 10^{0.1} - 1 = 0.2589$$

$$\omega_s^{14} = 10^3 - 1 = 999$$

$$\frac{\omega_s}{\omega_p} = \sqrt[14]{\frac{999}{0.2589}} = 1.804$$

(b) For the Chebyshev characteristic $(\omega_p = 1)$

$$10\log(1 + \epsilon^2) = 1 \quad \Longrightarrow \quad \epsilon^2 = 0.2589$$

$$10\log[1 + \epsilon^2 C_7^2(\omega_s)] = 30 \quad \Longrightarrow \quad \epsilon^2 C_7^2(\omega_s) = 999$$

$$C_7^2(\omega_s) = \frac{999}{0.2589} = 3858.3 \quad \Longrightarrow \quad C_7(\omega_s) = 62.115$$

$$\omega_s = \cosh\left[\frac{1}{7}\cosh^{-1} 62.115\right] = 1.247$$

(c) For the elliptic-function characteristic, from Table A.15, we find

$$\omega_s/\omega_p = 1.015$$

EXAMPLE 2.5. For $n = 5$, $\alpha_p = 0.5$ dB, and $\omega_s/\omega_p = 1.5$, determine α_s for the three types of magnitude characteristics.

SOLUTION   (a) For the Butterworth characteristic

$$\omega_p^{10} = 10^{0.05} - 1 = 0.1220 \quad \Longrightarrow \quad \omega_p = 0.8103$$

$$\alpha_s = 10\log[1 + (1.5 \times 0.8103)^{10}] = 9.0505 \text{ dB}$$

(b) For the Chebyshev characteristic ($\omega_p = 1$)

$$\epsilon^2 = 10^{0.05} - 1 = 0.1220$$

$$\alpha_s = 10\log[1 + \epsilon^2 C_5^2(1.5)] = 26.65 \text{ dB}$$

(c) For the elliptic-function characteristic, from Table A.17, we find

$$\alpha_s = 50.61 \text{ dB}$$

From these examples, it is seen that the improvements from the Butterworth characteristic, to the Chebyshev characteristic, and to the elliptic-function characteristic are quite dramatic. If a filtering requirement is demanding and the magnitude is the major consideration, the elliptic-function characteristic should be the first choice. However, as we shall see in some of the examples in Chapter 3, as the magnitude characteristics improve, the phase functions, or equivalently the delay functions, become progressively worse. So we don't get something for nothing.

2.6   Summary

In this chapter, we have introduced several standard lowpass characteristics that are in common use as the magnitudes of lowpass filters. These include the Butterworth, the Chebyshev and its variations, and the elliptic-function characteristics. Their relative merits are briefly compared.

In the next chapter, we shall deal with the network functions that have the magnitude characteristics described in this chapter.

Problems

2.1 Determine the lowest n of a Butterworth lowpass magnitude characteristic such that

$$\alpha_p = 0.5 \text{ dB} \qquad \alpha_s \geq 30 \text{ dB} \qquad \omega_s/\omega_p = 1.3$$

Also determine the exact value of α_s and the half-power ω.

2.2 Determine the lowest n of a Butterworth lowpass magnitude characteristic such that

$$\alpha_p = 1 \text{ dB} \qquad \alpha_s \geq 45 \text{ dB} \qquad \omega_s/\omega_p = 1.75$$

Also determine the exact value of α_s.

2.3 Determine the lowest n of a Butterworth lowpass magnitude characteristic such that

$$\alpha_p \leq 0.75 \text{ dB} \qquad \alpha_s = 30 \text{ dB} \qquad \omega_s/\omega_p = 1.5$$

Also determine the exact value of α_p. What is the value of ω at which the attenuation is 0.75 dB?

2.4 Determine the lowest n of a Butterworth lowpass magnitude characteristic such that

$$\alpha_p = 1.5 \text{ dB} \qquad \alpha_s = 40 \text{ dB} \qquad \omega_s/\omega_p \leq 1.3$$

Also determine the exact value of ω_s/ω_p.

2.5 It is desired to use a seventh-order Butterworth lowpass characteristic with $\alpha_p = 1$ dB. What would be the exact attenuation at $\omega = 2.5\omega_p$?

2.6 Use Euler's formula and the definition of the hyperbolic cosine to show that

$$\cos(n \cos^{-1} \omega) \equiv \cosh(n \cosh^{-1} \omega)$$

2.7 Determine the values of ω in the range $0 < \omega < 1$ where the Chebyshev lowpass characteristic of the fourth order has either a maximum or a minimum.

2.8 Determine the attenuation at $\omega = 3$ for a Chebyshev lowpass characteristic with $\alpha_p = 1.25$ dB and $n = 15$.

2.9 Determine the lowest n of a Chebyshev lowpass characteristic such that

$$|H(j\omega)|_{\max} = 1 \qquad |H(j\omega)|_{\min} = 0.95 \qquad \text{for } 0 < f < 1 \text{ kHz}$$

$$|H(j\omega)| \leq 0.05 \qquad \text{for } f > 1.5 \text{ kHz}$$

2.10 Determine the value of ϵ and the lowest n of a Chebyshev lowpass characteristic such that

$$\alpha_p = 1.75 \text{ dB} \qquad \alpha_s \geq 38 \text{ dB} \qquad \omega_s/\omega_p = 1.4$$

What is the exact value of the attenuation at $\omega = 1.4$?

2.11 Determine the lowest n of a Chebyshev lowpass characteristic such that

$$\alpha_p = 1.5 \text{ dB} \qquad \alpha_s \geq 30 \text{ dB} \qquad \omega_s \leq 1.5$$

What is the exact attenuation at $\omega = 1.5$? What is the value of ω at which the attenuation is exactly 30 dB?

2.12 Determine the lowest n of a Chebyshev lowpass characteristic such that

$$\alpha_p \leq 0.1 \text{ dB} \qquad \alpha_s = 30 \text{ dB} \qquad \omega_s = 1.2$$

What is the exact value of α_p?

2.13 For the magnitude function

$$|H(j\omega)| = \frac{0.5}{\sqrt{1 + 0.3^2 C_6^2(\omega)}}$$

What are the maximum and minimum gains in the pass band $(0 < \omega < 1)$? What is the attenuation at $\omega = 2$ compared with the maximum gain?

2.14 Prove that another recursive formula for Chebyshev polynomials is

$$C_{2n}(\omega) = 2C_n^2(\omega) - 1$$

2.15 It is known that if a lowpass elliptic-function filter is used for

$$\alpha_p = 0.5 \text{ dB} \qquad \alpha_s = 45 \text{ dB} \qquad \omega_s/\omega_p = 1.2$$

it is necessary that $n = 6$. What is the necessary n if a Butterworth filter is used? What if a Chebyshev filter is used?

2.16 Show that for $\omega > 1$, the Chebyshev polynomial may be alternatively expressed as

$$C_n(\omega) = \frac{1}{2} \left\{ \left[\omega + (\omega^2 - 1)^{\frac{1}{2}}\right]^n + \left[\omega + (\omega^2 - 1)^{\frac{1}{2}}\right]^{-n} \right\}$$

or

$$C_n(\omega) = \frac{1}{2} \left\{ \left[\omega - (\omega^2 - 1)^{\frac{1}{2}}\right]^n + \left[\omega - (\omega^2 - 1)^{\frac{1}{2}}\right]^{-n} \right\}$$

2.17 Show that the inverse Chebyshev lowpass characteristic of Fig. 2.8 for which

$$|H_c(j\omega)|^2 = \frac{\epsilon^2 C_n^2(1/\omega)}{1 + \epsilon^2 C_n^2(1/\omega)}$$

is maximally flat at $\omega = 0$.

2.18 Show that for the $R_N(\omega)$ defined in (2.29) and (2.30),

$$R_N\left(\frac{1}{\omega}\right) = \frac{1}{R_N(\omega)}$$

2.19 For the $|H(j\omega)|^2$ defined in (2.31), show that if

$$A_1^2 = \frac{1}{1 + \epsilon^2 A^2}$$

then

$$A_2^2 = \frac{1}{1 + \epsilon^2 (1/A^2)}$$

2.20 In (2.31), we let $\epsilon = 0.4$, $\omega_1 = 0.9$, $N = 1$, and use the $R_N(\omega)$ of (2.29). In other words,

$$|H(j\omega)|^2 = \frac{1}{1 + 0.16 \left[\dfrac{0.81 - \omega^2}{1 - 0.81\omega^2}\right]^2}$$

and $|H(j\omega)|$ will vary as in the plot shown below. Determine A_1,

A_2, ω_p, and ω_s.

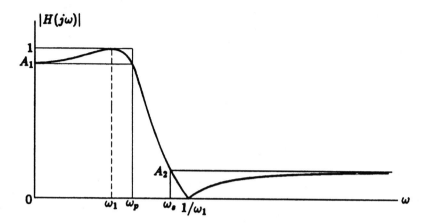

2.21 In (2.31), we let $\varepsilon = 0.5$, $\omega_1 = \sqrt{0.9}$, $N = 1$, and use the $R_N(\omega)$ of (2.30). In other words,

$$|H(j\omega)|^2 = \cfrac{1}{1 + 0.25 \left[\cfrac{\omega(0.9 - \omega^2)}{1 - 0.9\omega^2}\right]^2}$$

and $|H(j\omega)|$ will vary as shown below. First, determine the value of ω_a at which $|H(j\omega)|$ is a minimum in the pass band. Then determine A_1, A_2, ω_p, and ω_s.

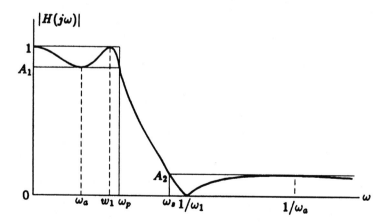

Chapter 3

Network functions

Once a magnitude characteristic has been chosen for a particular filter application, the next step is to determine a network function that not only has this magnitude characteristic, but is also realizable. By realizable, we mean that the function must be such that a workable network at least exists in theory, can be implemented with real-world components, and, barring unrealistic assumptions on these components, can be constructed and expected to perform the task accordingly. For example, the network function must be such that it has no pole in the right half of the s plane. If it does, the network will not be stable. The constructed network will either become nonlinear and function improperly or self destruct.

3.1 General procedure

We now outline the procedure that will enable us to obtain a network function $H(s)$ when its $|H(j\omega)|$ (or equivalently, $|H(j\omega)|^2$) is given. In (2.1), we stated that for a given $H(s)$ its squared magnitude on the $j\omega$ axis can be obtained by

$$|H(j\omega)|^2 = H(s)H(-s)\Big|_{s=j\omega} \tag{3.1}$$

Our current objective is just the opposite. Here, we have an $|H(j\omega)|^2$ given and we write

$$H(s)H(-s) = |H(j\omega)|^2_{\omega^2=-s^2} = \frac{A(\omega^2)}{B(\omega^2)}\bigg|_{\omega^2=-s^2} \tag{3.2}$$

If we let

$$H(s) = \frac{P(s)}{Q(s)} \tag{3.3}$$

then we can write

$$P(s)P(-s) = A(\omega^2)\Big|_{\omega^2=-s^2} \quad \text{and} \quad Q(s)Q(-s) = B(\omega^2)\Big|_{\omega^2=-s^2} \tag{3.4}$$

The procedure for obtaining a $P(s)$ for a given $A(\omega^2)$ and that for obtaining a $Q(s)$ for a given $B(\omega^2)$ are mathematically identical. We shall arbitrarily choose to deal with $A(\omega^2)$ and $P(s)$.

Since $P(s)$ is a polynomial with real coefficients, its zeros must be either real or occur in conjugate pairs. The zeros of $P(-s)$ are the negative of those of $P(s)$. Hence the zeros of $P(s)P(-s)$ can occur only in groups each of which is one of the three types shown in Fig. 3.1.

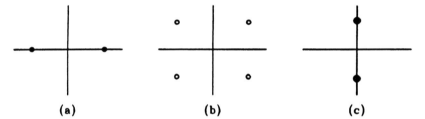

(a)   (b)   (c)

Figure 3.1: Patterns of zeros of $P(s)P(-s)$.

In group type (a), the zeros are real. For each zero in one halfplane, there must also be another one in the other halfplane. Group type (b) contains complex zeros. They must occur in sets of four symmetric about both axes. We describe this type of group as possessing *quadrantal symmetry*. Group type (c) are j-axis zeros. In addition to being conjugate pairs, zeros of this type must be of even multiplicity. In Fig. 3.1, a conjugate pair of double zeros is indicated.

In obtaining a $P(s)$ for a given $A(\omega^2)$, we first replace every ω^2 in $A(\omega^2)$ with $-s^2$, or every ω with s/j. Then $A(-s^2)$ is factored to reveal all its zeros. For real zeros, the corresponding factors will appear as $(s + a)(s - a)$. For complex zeros, the corresponding factors will appear as $(s^2 + as + b)(s^2 - as + b)$. For j-axis zeros, the factors will appear as $(s^2 + \omega_i^2)^2$.

The construction of $P(s)$ is now clear. For each type (a) or (b) group of zeros, we allot those in either halfplane to $P(s)$, and those in the other halfplane to $P(-s)$. As far as $A(\omega^2)$ is concerned, which halfplane zeros are allotted to $P(s)$ doesn't matter. For a type (c) group of zeros, we allot one half of each multiple zero to $P(s)$ and $P(-s)$.

For example, if

$$A(-s^2) = (s+2)(s-2)(s^2+2s+5)(s^2-2s+5)(s^2+6)^2 \qquad (3.5)$$

$P(s)$ could be any of the following.

$$(s+2)(s^2+2s+5)(s^2+6)$$

$$(s-2)(s^2+2s+5)(s^2+6)$$

$$(s+2)(s^2-2s+5)(s^2+6)$$

$$(s-2)(s^2-2s+5)(s^2+6)$$

Obviously, for each of the four choices, the resultant expressions $P(s)P(-s)$ are exactly alike.

The procedure for obtaining a $Q(s)$ for a given $B(\omega^2)$ is exactly the same as that for obtaining a $P(s)$ for a given $A(\omega^2)$. However, since $Q(s)$ must be Hurwitz,[1] we must always choose the left-halfplane zeros from each group. Hence, if (3.5) were a $B(-s^2)$, then the only choice for $Q(s)$ would be $(s+2)(s^2+2s+5)(s^2+6)$.

The various ways in which the zeros of $A(-s^2)$ from each group are allotted to $P(s)$ do not affect $A(\omega^2)$. They do have effects on the phase function of the resultant $H(s)$. When only zeros in the left halfplane are included in $P(s)$, the $H(s)$ will have the lowest phase at each frequency. The network function so formed is called the *minimum-phase* function. If right-halfplane zeros are included in $P(s)$, the network function is a *nonminimum-phase* function.

EXAMPLE 3.1. Obtain the network function whose magnitude squared is

---

[1] A polynomial is Hurwitz if its zeros are all in the open left halfplane. This definition does not allow zeros to lie on the imaginary axis. In some limiting situations, simple zeros are allowed to lie on the imaginary axis. In such situations, the polynomials are no longer, strictly speaking, Hurwitz. Polynomials with no right-halfplane zeros and with simple zeros on the imaginary axis are sometimes referred to as *modified Hurwitz polynomials*.

$$|H(j\omega)|^2 = \frac{\omega^4 + 1}{\omega^2(\omega^4 + 6\omega^2 + 25)}$$

SOLUTION   We have

$$H(s)H(-s) = \frac{s^4 + 1}{-s^2(s^4 - 6s^2 + 25)}$$

$$= \frac{(s - \underline{/45°})(s - \underline{/-45°})(s - \underline{/135°})(s - \underline{/-135°})}{-s^2(s + 2 + j)(s + 2 - j)(s - 2 + j)(s - 2 - j)}$$

Hence we can have either

$$H(s) = \frac{(s + 0.707107 + j0.707107)(s + 0.707107 - j0.707107)}{s(s + 2 - j)(s + 2 + j)}$$

$$= \frac{s^2 + 1.41421s + 1}{s(s^2 + 4s + 5)}$$

or

$$H(s) = \frac{(s - 0.707107 + j0.707107)(s - 0.707107 - j0.707107)}{s(s + 2 - j)(s + 2 + j)}$$

$$= \frac{s^2 - 1.41421s + 1}{s(s^2 + 4s + 5)}$$

EXAMPLE 3.2. Obtain the network function whose magnitude squared is

$$|H(j\omega)|^2 = \frac{\omega^2 + 9}{\omega^6 - 3\omega^4 + 12\omega^2 + 100}$$

SOLUTION   We now have

$$H(s)H(-s) = \frac{s^2 - 9}{s^6 + 3s^4 + 12s^2 - 100}$$

$$= [(s + 3)(s - 3)]/[(s - 1.78026)(s + 1.78026)(s + 1.12528 + j2.08588)$$

$$(s+1.12528-j2.08588)(s-1.12528+j2.08588)(s-1.12528-j2.08588)]$$

Thus

$$H(s) = \frac{s \pm 3}{(s + 1.78026)(s + 1.12528 + j2.08588)(s + 1.12528 - j2.08588)}$$

$$= \frac{s \pm 3}{(s + 1.78026)(s^2 + 2.25056s + 5.61716)}$$

$$= \frac{s \pm 3}{s^3 + 4.03081s^2 + 9.62374s + 10}$$

The computational steps to obtain the denominator of $H(s)$ can be carried out using MATLAB. The following are the commands and results.

```
% Specify denominator coefficient vector
» a = [1 0 3 0 12 0 -100];
% Obtain denominator zeros
» b = roots(a)
b =
-1.1253 + 2.0859i
-1.1253 - 2.0859i
 1.1253 + 2.0859i
 1.1253 - 2.0859i
 1.7803
-1.7803
% Form the denominator polynomial with the three
% left-halfplane zeros
» c = poly([b(1) b(2) b(6)])
c = 1 4.0308 9.6237 10
```

3.2   Network functions for Butterworth filters

For a network function whose magnitude is the Butterworth normalized lowpass characteristic, we have

$$H(s)H(-s) = \frac{1}{1 + (-s^2)^n} \tag{3.6}$$

We only have to deal with $B(-s^2)$. From (3.6), the zeros of $Q(s)Q(-s)$ are the roots of the equation

$$1 + (-s^2)^n = 0 \qquad \text{or} \qquad (-s^2)^n = -1 = e^{j(\pi + 2k\pi)} \tag{3.7}$$

where k is any integer. Equation (3.7) leads to

$$-s^2 = e^{j\frac{(2k+1)}{n}\pi} \quad \text{or} \quad s^2 = e^{j\left[\frac{(2k+1)}{n}\pi - \pi\right]} \tag{3.8}$$

Hence the $2n$ zeros of $Q(s)Q(-s)$ are

$$s_k = e^{j\left[\frac{(2k+1)}{2n}\pi - \frac{\pi}{2}\right]} \tag{3.9}$$

Alternately, if we write $s_k = \sigma_k + j\omega_k$, we have

$$\sigma_k = \cos\left[\frac{(2k+1)}{2n}\pi - \frac{\pi}{2}\right] = \sin\left[\frac{(2k+1)}{2n}\pi\right]$$

$$\omega_k = \sin\left[\frac{(2k+1)}{2n}\pi - \frac{\pi}{2}\right] = -\cos\left[\frac{(2k+1)}{2n}\pi\right]$$

Equation (3.9) indicates that the $2n$ zeros of $Q(s)Q(-s)$ are uniformly spaced around a unit circle whose angles are $360°/2n$ apart. If n is odd, two of the zeros will lie on the real axis. If n is even, all the zeros are complex conjugate pairs, and thus there will be none on the real axis. Figure 3.2 shows the zero distribution of $Q(s)Q(-s)$ for $n = 6$ and $n = 9$.

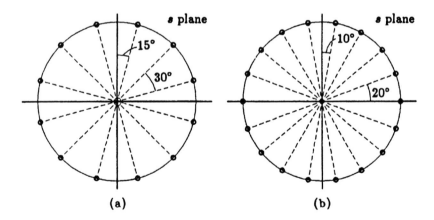

Figure 3.2: Zeros of $Q(s)Q(-s)$ for (a) $n = 6$ and (b) $n = 9$.

Since $Q(s)$ is the denominator of $H(s)$, it must be Hurwitz. Hence we must include only left-halfplane zeros of $B(-s^2)$ to form $Q(s)$. Polynomials formed to have left-halfplane zeros are known as *Butterworth polynomials*. Of course, zeros of $Q(s)$ are poles of $H(s)$.

EXAMPLE 3.3. Obtain the network function of the normalized Butterworth fourth-order lowpass filter.

SOLUTION   The poles of $H(s)$ are shown in Fig. 3.3.

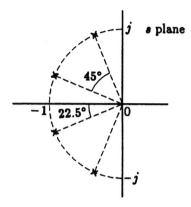

Figure 3.3: Poles of the fourth-order Butterworth lowpass filter.

Hence

$$H(s) = \frac{1}{(s - \underline{/112.5°})(s - \underline{/-112.5°})(s - \underline{/157.5°})(s - \underline{/-157.5°})}$$

$$= \frac{1}{(s + 0.382683 \pm j0.923880)(s + 0.923880 \pm j0.382683)}$$

$$= \frac{1}{(s^2 + 0.765367s + 1)(s^2 + 1.84778s + 1)}$$

$$= \frac{1}{s^4 + 2.61313s^3 + 3.41421s^2 + 2.61313s + 1}$$

This example can also be worked using MATLAB. Following the pattern of steps in Example 2, we have the following steps.

```
≫ a = [1 zeros(1,7) 1];
≫ b = roots(a)
b =
-0.9239 + 0.3827i
-0.9239 - 0.3827i
-0.3827 + 0.9239i
-0.3827 - 0.9239i
```

```
    0.3827 + 0.9239i
    0.3827 - 0.9239i
    0.9239 + 0.3827i
    0.9239 - 0.3827i
» c = poly([b(1:4)])
c = 1 2.6131 3.4142 2.6131 1
```

EXAMPLE 3.4. Obtain the network function of the normalized fifth-order Butterworth lowpass filter.

SOLUTION   The poles of $H(s)$ are shown in Fig. 3.4. Hence

$$H(s) = \frac{1}{(s+1)(s - \underline{/108°})(s - \underline{/-108°})(s - \underline{/144°})(s - \underline{/-144°})}$$

$$= \frac{1}{(s+1)(s+0.309017 \pm j0.951057)(s+0.809017 \pm j0.587785)}$$

$$= \frac{1}{(s+1)(s^2+0.618034s+1)(s^2+1.61803s+1)}$$

$$= \frac{1}{s^5 + 3.23607s^4 + 5.23607s^3 + 5.23607s^2 + 3.23607s + 1}$$

Using MATLAB, we have the following steps.

```
» a = [1 zeros(1,9) -1];
» b = roots(a)
b =
-1.0000
-0.8090 + 0.5878i
-0.8090 - 0.5878i
-0.3090 + 0.9511i
-0.3090 - 0.9511i
   0.3090 + 0.9511i
   0.3090 - 0.9511i
   1.0000
   0.8090 + 0.5878i
   0.8090 - 0.5878i
» c = poly([b(1:5)])
c = 1 3.2361 5.2361 5.2361 3.2361 1
```

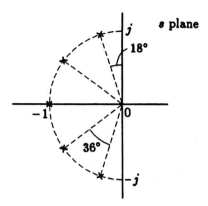

Figure 3.4: Poles of the fifth-order Butterworth lowpass filter.

It might be well to point out that normalized Butterworth polynomials have two interesting features. One of them is that the constant term is always unity. The other is that the coefficients of the polynomial are always symmetric.

A number of Butterworth polynomials are tabulated in Appendix A. Coefficients of these polynomials can also be obtained using the command buttap of MATLAB. This command has the format

$$[z,p,k] = \text{buttap}(n)$$

where n is the order of the filter function, z is the vector containing the zeros (it will always be empty), p is the vector containing the poles, and k is the proportionality constant (it will always be unity). Hence the Butterworth polynomials can be formed by using the poles returned by this command. The following are the steps by which the results of Examples 2 and 3 can also be obtained.

```
% Obtain the data for the 4th-order Butterworth function
» [z p k] = buttap(4);
% Obtain the denominator polynomial coefficients from
% the poles
» c = poly(p)
```

```
c = 1 2.6131 3.4142 2.6131 1

% Obtain the data for the 5th-order Butterworth function
» [z p k] = buttap(5);
» d = poly(p)
d = 1 3.2361 5.2361 5.2361 3.2361 1
```

3.3   Network functions for Chebyshev filters

For a Chebyshev lowpass magnitude characteristic, from (2.22) we have

$$H(s)H(-s) = \frac{K^2}{Q(s)Q(-s)} = \frac{K^2}{1 + \epsilon^2 C_n^2(-js)}$$

Again we have to deal with only the denominator polynomials. The zeros of $Q(s)Q(-s)$ are the roots of the equation

$$1 + \epsilon^2 C_n^2(-js) = 0$$

which gives

$$C_n(-js) = \cos[n\cos^{-1}(-js)] = \pm j\frac{1}{\epsilon} \tag{3.10}$$

We let

$$\cos^{-1}(-js) = u + jv \tag{3.11}$$

Alternatively,

$$-js = \cos(u + jv) = \cos u \cosh v - j \sin u \sinh v$$

or

$$s = \sin u \sinh v + j \cos u \cosh v \tag{3.12}$$

Equation (3.10) becomes

$$\cos[n(u + jv)] = \cos nu \cosh nv - j \sin nu \sinh nv = \pm j\frac{1}{\epsilon} \tag{3.13}$$

Equating real and imaginary parts, we obtain

$$\cos nu \cosh nv = 0 \tag{3.14}$$

$$-\sin nu \sinh nv = \pm\frac{1}{\epsilon} \tag{3.15}$$

Since for real v, $\cosh nv > 1$, we must have $\cos nu = 0$ or nu must be an odd multiple of $\pi/2$. Thus

$$u = \frac{(2k-1)\pi}{2n} \tag{3.16}$$

where k may be any integer. Thus $\sin nu = \pm 1$ and[2]

$$\sinh nv = \frac{1}{\epsilon} \quad \text{and} \quad v = \frac{1}{n}\sinh^{-1}\left[\frac{1}{\epsilon}\right] \tag{3.17}$$

Hence the locations of the zeros are given by

$$s_k = \sin\left[\frac{(2k-1)\pi}{2n}\right]\sinh v + j\cos\left[\frac{(2k-1)\pi}{2n}\right]\cosh v \tag{3.18}$$

In summary, once ϵ and n are known, v can be determined using (3.17) and the zeros can be calculated using (3.18). Although k may be any integer, there are only $2n$ distinct zeros given by (3.18).

If we let $s_k = \sigma_k + j\omega_k$, then

$$\sigma_k = \sin\left[\frac{(2k-1)\pi}{2n}\right]\sinh v \tag{3.19}$$

$$\omega_k = \cos\left[\frac{(2k-1)\pi}{2n}\right]\cosh v \tag{3.20}$$

From (3.19) and (3.20) it is seen that

$$\frac{\sigma_k^2}{\sinh^2 v} + \frac{\omega_k^2}{\cosh^2 v} = 1 \tag{3.21}$$

Thus all zeros of $Q(s)Q(-s)$ lie on an ellipse with a major semiaxis (on the imaginary axis) equal to $\cosh v$ and a minor semiaxis (on the real axis) equal to $\sinh v$. Figure 3.5 shows how one of the zeros can

---

[2]Below, we arbitrarily choose both sides to be positive for convenience. As the values of k are varied sufficiently, the other sign choices would also be included eventually.

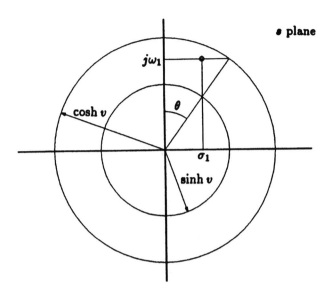

Figure 3.5: Geometry of the location of a pole of Chebyshev filter.

be located. Two circles are drawn with their radii equal to $\cosh v$ and $\sinh v$ respectively. The angle θ is equal to $(2k-1)\pi/2n$. For $k = 1$, σ_1 and ω_1 have their values shown on the two axes. Hence this particular zero is located as shown in the figure. Other zeros can be located in a similar manner by varying the radial lines as k is varied.

EXAMPLE 3.5. Obtain the fourth-order network function of a low-pass Chebyshev filter with $\alpha_p = 0.75$ dB. Determine K such that the maximum value of $|H(j\omega)|$ is unity.

SOLUTION   We have

$$10\log(1 + \epsilon^2) = 0.75$$

Solving

$$\epsilon^2 = 0.188502 \qquad \Longrightarrow \qquad \epsilon = 0.434168$$

From (3.17)

$$v = \frac{1}{4}\sinh^{-1}\left(\frac{1}{\epsilon}\right) = 0.392894$$

$$\sinh v = 0.403080 \qquad \text{and} \qquad \cosh v = 1.07818$$

Thus

$$s_{1,2} = -\sin\frac{\pi}{8}\sinh v \pm j\cos\frac{\pi}{8}\cosh v = -0.154252 \pm j0.996109$$

$$s_{3,4} = -\sin\frac{3\pi}{8}\sinh v \pm j\cos\frac{3\pi}{8}\cosh v = -0.372398 \pm j0.412602$$

These poles are shown to scale in Fig. 3.6. A small amount of algebraic

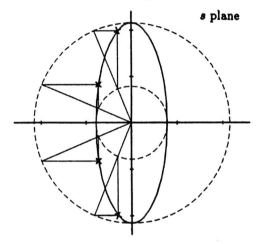

s plane

Figure 3.6: Poles of a fourth-order Chebyshev filter function.

work will yield

$$H(s) = \frac{K}{s^4 + 1.05330s^3 + 1.55472s^2 + 0.852035s + 0.313871}$$

To make $|H(j\omega)|_{\max} = 1$ and since we know that $H(0) = 1/\sqrt{1+\epsilon^2}$, we must make

$$\frac{K}{0.313871} = \frac{1}{\sqrt{1+0.188502}}$$

Hence

$$K = 0.287906$$

A number of normalized Chebyshev lowpass functions are tabulated in Appendix A. These functions, as well as characteristics not listed in Appendix A, can be obtained by using the command cheb1ap of MATLAB. This command has the format

$$[z,p,k] = \text{cheb1ap}(n,Rp)$$

where n is the order of the filter function, Rp is the passband ripple in dB (or α_p), z is the vector containing the zeros (it will always be empty), p is the vector containing the poles, and k is the proportionality constant adjusted such that $|H(j\omega)|_{\max} = 1$. The results of Example 3.5 can be obtained using the following steps.

```
% Obtain the data of a 4th-order lowpass Chebyshev
% network function
> [z p k] = cheb1ap(4,0.75);
% Obtain the polynomial coefficients from the poles
> e = poly(p)
e = 1 1.0533 1.5547 0.8520 0.3139
k = 0.2879
```

3.4 Network functions for elliptic-function filters

As was mentioned in Section 2.4, magnitude characteristics can be constructed such that equal-ripple variation occurs in both the pass band and the stop band. It was also remarked there that network functions for this type of filters can be obtained without first obtaining their magnitude characteristics. We shall indicate qualitatively one method by which this can be accomplished. This method makes use of the conformal transformation [Gr]. Conformal transformation maps points in one complex plane onto another while preserving all angular relationships.[3]

Specifically, the transformation

$$s = -j\,\text{sn}(jp, k) \tag{3.22}$$

maps the entire s plane onto a rectangle in the p plane as depicted in Fig. 3.7. The function sn is the Jacobian elliptic sine function, k is the modulus, K is the complete elliptic integral of modulus k, and K' is the complete elliptic integral of modulus $k' = \sqrt{1 - k^2}$.

---

[3]The reader may be familiar with the Smith chart used as an aid to the study of transmission lines and microwave circuits. The Smith chart is simply the mapping of the right half of a complex plane onto a circle using conformal transformation.

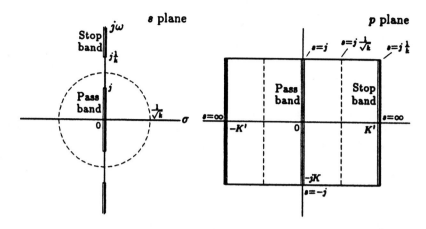

Figure 3.7: The mapping of the s plane onto a rectangle.

In our application, we consider $s = j\omega$, $0 < \omega < 1$, to be the pass band and $s = j\omega$, $1/k < \omega < \infty$, to be the stop band. These regions in the s plane and their mapping in the p plane are indicated in Fig. 3.7. The usual symmetry of any $H(s)H(-s)$ about both axes in both planes is implied.

The rectangle in the p plane of Fig. 3.7 is only one *cell* of an infinite array of the same rectangle repeated in both the horizonal and vertical directions. This is because elliptic functions are doubly periodic. Transform (3.22) actually maps the s plane onto an infinite number of rectangles bordering one another in a regular fashion as shown in Fig. 3.8. The rectangular cell centered at the origin is shown with slightly heavier borders. Now if we place poles and zeros in the p plane appropriately, the desired magnitude variation will result. We cannot have poles on the j axis, so they are placed some distance away from the j axis. We wish to place the zeros in the stop band, so double zeros are placed along the vertical lines K' away from the j axis. This pattern is repeated every $2K'$ horizontally. Vertically, poles and zeros are uniformly spaced. Because of the cyclic locations of poles and zeros, a network function having these poles and zeros must vary cyclically with equal maxima and equal minima along any vertical line. Hence the equal-ripple variation desired for the elliptic filters is automatically produced this way.

The pole-zero pattern of Fig. 3.8 corresponds to a fifth-order elliptic filter. The poles and zeros of $H(s)H(-s)$ are shown in Fig. 3.9. To obtain $H(s)$ we simply drop the right-halfplane poles and keep one of each double zero on the j axis.

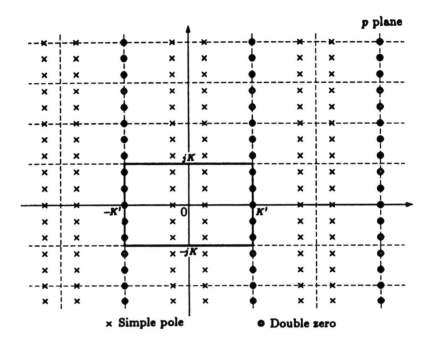

Figure 3.8: The rectangle of Fig. 3.7 is repeated in both directions in the p plane.

Referring to Fig. 3.7, we see that network functions obtained this way have the pass band normalized to $\omega_p = 1$. The modulus k controls the relative lengths of the two sides of the rectangle. Thus k directly determines the transition-band ratio as $\omega_s/\omega_p = \omega_s = 1/k$. The distance of the columns of poles from the j axis determines the passband ripple α_p. The closer these poles are to the j axis, the higher the passband ripple will be.

Since most engineers are not familiar with elliptic functions, other derivations have been developed to achieve the same result. However, these other derivations are generally not any simpler, although they may circumvent the need to use elliptic functions. A quantitative development of how elliptic-filter functions can be obtained is beyond the scope of this text. As is the case in practice, we shall rely on precalculated tables or computer software when network functions for elliptic filters are needed. One easy-to-use set of tables has been compiled by the author [Su3]. Also, several selected tables are given in Appendix A.

Elliptic-filter functions can also be obtained by using MATLAB. The

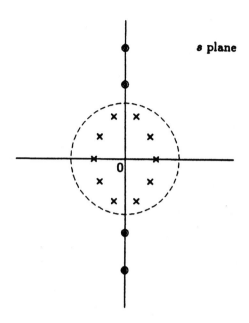

Figure 3.9: Pole-zero locations of $H(s)H(-s)$ of a fifth-order elliptic filter.

program gives the necessary data for a given order, a given passband ripple, and a given stopband ripple. The format of the command is

$$[z,p,k] = ellipap(n,Rp,Rs)$$

where n is the order of the filter function, Rp is the passband ripple in dB (or α_p), Rs is the stopband attenuation in dB (or α_s), z is the vector containing the zeros, p is the vector containing the poles, and k is the proportionality constant adjusted such that $|H(j\omega)|_{max} = 1$.

EXAMPLE 3.6. Use MATLAB to obtain the network function of a fifth- order elliptic filter with $\alpha_p = 1$ dB and $\alpha_s = 30$ dB.

SOLUTION We start with the command

```
» [z,p,k] = ellipap(5,1,30)
z = -1.4688i, +1.4688i, -1.1171i, +1.1171i
p = -0.2009 - 0.8085i, -0.2009 + 0.8085i,
    -0.0338 - 1.0003i, -0.0338 + 1.0003i
    -0.4506
k = 0.1164
```

To form the numerator factors, we convert one pair of conjugate zeros at a time into a polynomial.

```
> n1 = poly([z(1) z(2)])
n1 = 1 0 2.1574
> n2 = poly([z(3) z(4)])
n2 = 1 0 1.2480
```

Similarly, for the denominator,

```
> d1 = poly([p(1) p(2)])
d1 = 1 0.4018 0.6940
> d2 = poly([p(3) p(4)])
d2 = 1 0.0676 1.0018
```

Hence the network function is

$$H(s) = \frac{0.1164(s^2 + 2.1574)(s^2 + 1.2480)}{(s + 0.4506)(s^2 + 0.4018s + 0.6940)(s^2 + 0.0676s + 1.0018)}$$

which is the same as the corresponding entry in Table A.15.

3.5 Bessel-Thomson filter functions

So far, our focus has been on the magnitudes of filter functions. There are applications in which the phase of a filter function defined in (1.2) is important. This is because the phase function is directly related to the group delay, as expressed in (1.4), of the signal. Group delay is of particular importance when signal waveshapes are an important consideration. For example, it is generally accepted that, for audio applications, the different delays for signals of different frequencies are not noticeable to human ears. On the other hand, in the transmission of video signals, if the delay is not substantially constant throughout the pass band, the picture will be distorted.

One standard lowpass filter that focuses on the delay of the filter at low frequencies was first developed by Thomson [Thn]. Thomson obtained the lowpass functions such that the delay is maximally flat at the origin. The approach he used can be illustrated by the second-order function. We let

$$H_2(s) = \frac{b_0}{s^2 + b_1 s + b_0} \tag{3.23}$$

The phase function of $H(s)$ is

$$\phi(\omega) = -\tan^{-1}\left[\frac{b_1\omega}{b_0 - \omega^2}\right] \tag{3.24}$$

The group delay is

$$T_d(\omega) = -\frac{d\phi}{d\omega} = \frac{b_1\omega^2 + b_1 b_0}{\omega^4 + (b_1^2 - 2b_0)\omega^2 + b_0^2} \tag{3.25}$$

To normalize the value of T_d at $\omega = 0$ to unity, we make the constant terms equal. Or

$$b_1 b_0 = b_0^2 \quad \Longrightarrow \quad b_1 = b_0 \tag{3.26}$$

To achieve the maximally flat delay at the origin, we set the coefficients of ω^2 in the numerator and the denominator of (3.25) equal to each other. This requires that

$$b_1 = b_1^2 - 2b_0 \tag{3.27}$$

The combination of (3.26) and (3.27) leads to $b_1 = b_0 = 3$. Hence, for $T_d(\omega)$ to be maximally flat at the origin with a delay of 1 second, we have

$$H(s) = \frac{3}{s^2 + 3s + 3} \tag{3.28}$$

This process, although routine, is tedious. We shall follow a process due to Storch [Sto], which is a great deal more elegant and less brute-force. To achieve a distortionless transmission, it is necessary that the impulse response of the system be

$$h(t) = \delta(t - \tau) \tag{3.29}$$

where $\delta(t)$ is the unit impulse function and τ is the group delay. In the frequency domain, we require

$$H(s) = e^{-s} \tag{3.30}$$

in which we have normalized the delay to 1 second. Equation (3.30) may alternatively be expressed as

$$H(s) = \frac{1}{e^s} = \frac{1}{\sinh s + \cosh s} \tag{3.31}$$

If we use an all-pole rational function[4] to approximate this function, we will have

---

[4]An all-pole function is one that has only finite poles, no finite zeros. Butterworth and Chebyshev lowpass filter functions are examples of all-pole functions.

$$H_n(s) = \frac{K}{M(s) + N(s)} \approx H(s)$$

where $M(s)$ and $N(s)$ are the even and odd parts of the denominator polynomial. Then we can have

$$\frac{M(s)}{N(s)} \approx \frac{\cosh s}{\sinh s}$$

The series expansions of the hyperbolic functions are

$$\cosh s = 1 + \frac{s^2}{2!} + \frac{s^4}{4!} + \frac{s^6}{6!} + \cdots$$

$$\sinh s = s + \frac{s^3}{3!} + \frac{s^5}{5!} + \frac{s^7}{7!} + \cdots$$

By performing the long division, inverting, and repeating the steps, we can obtain the continued-fraction expansion

$$\frac{\cosh s}{\sinh s} = \frac{1}{s} + \cfrac{1}{\cfrac{3}{s} + \cfrac{1}{\cfrac{5}{s} + \cfrac{1}{\cfrac{7}{s} + \cdots}}} \tag{3.32}$$

To obtain the denominator of $H_n(s)$, we simply truncate the continued-fraction expansion after n terms, simplify the truncated expansion into a fraction, and add the numerator and the denominator of the fraction to get $M(s) + N(s)$. For example, for $n = 3$, we have

$$\frac{M(s)}{N(s)} = \frac{1}{s} + \cfrac{1}{\cfrac{3}{s} + \cfrac{1}{\cfrac{5}{s}}} = \frac{6s^2 + 15}{s^3 + 15s}$$

and

$$H_3(s) = \frac{15}{s^3 + 6s^2 + 15s + 15} \tag{3.33}$$

The numerator is made equal to the constant term of the denominator so the gain at low frequencies is normalized to unity. Storch was able to show that functions obtained this way have maximally flat delay at the origin and are identical to those obtained by Thomson.

We shall express $H_n(s)$ as

$$H_n(s) = \frac{b_0}{y_n(s)} = \frac{b_0}{s^n + b_{n-1}s^{n-1} + \cdots + b_1 s + b_0} \tag{3.34}$$

The denominator polynomial y_n is closely related to a class of Bessel polynomials. Its coefficients can be expressed in closed form in several ways [Su2]. One of them is

$$b_k = \frac{(2n-k)!}{2^{n-k}k!(n-k)!} \tag{3.35}$$

Polynomials $y_n(s)$ for n up to 10 are tabulated in Tables A.13 and A.14. The delay functions associated with $H_n(s)$ of (3.34) for n up to $n = 10$ are given in Fig. 3.10. Filters realizing this group of filter functions are

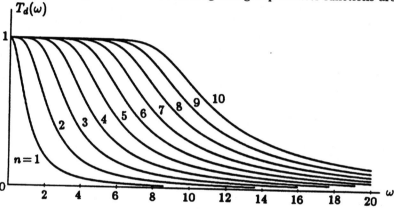

Figure 3.10: The maximally flat delay characteristics.

known as *Bessel filters*, *Thomson filters*, or *maximally flat delay filters*.

As a comparison of the magnitude and delay functions of this type of filters with those of the other three types of lowpass filters, we choose $n = 4$. Figure 3.11 shows four magnitude characteristics. One of them is the Bessel-Thomson filter. Another one is that of the normalized Butterworth filter with the 3 dB point located at $\omega = 1$. Still another one is that of the normalized Chebyshev filter with $\alpha_p = 1$ dB. The fourth one is that of the normalized elliptic filter with $\alpha_p = 1$ dB and $\alpha_s = 30$ dB.

From these characteristics, it is seen that the Bessel-Thomson magnitude characteristic is far inferior to those of the other three types. This should not be a surprise since the development of Bessel-Thomson filter functions is done solely with the constant delay as the objective. That this characteristic in Fig. 3.11 happens to be lowpass is really incidental.

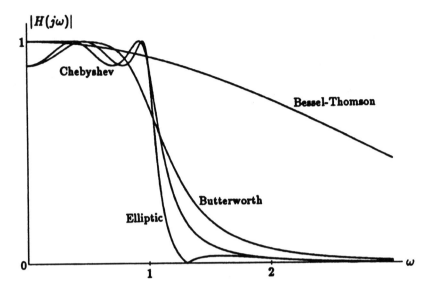

Figure 3.11: Magnitude characteristics of several fourth-order lowpass filters.

The function was derived with the assumption that it is all-pole and has the form of (3.34). As a consequence, this function has a generally lowpass magnitude characteristic simply because at high frequencies, the denominator tends to be large.

An allpass Bessel-Thomson function can be formed if we define

$$H_{AP,n}(s) = \frac{y_n(-s)}{y_n(s)} \tag{3.36}$$

The delay of this function is twice that given in (3.34).

The delay characteristics of these four types of lowpass filters are shown together in Fig. 3.12. These characteristics show that the delay functions associated with the three standard types of lowpass magnitude characteristics are quite variable. They also demonstrate that as the magnitude characteristics improve (from Bessel-Thomson to Butterworth to Chebyshev to elliptic), the delay characteristics become progressively more and more variable. Hence, the improvement of the standard lowpass characteristics from one type to another is not without costs. One such cost is the worsening of the delay characteristic as the magnitude characteristic improves.

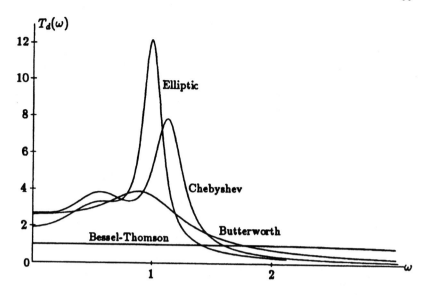

Figure 3.12: Delay characteristics of several fourth-order lowpass filters.

3.6 Delay equalization

As was mentioned in Chapter 1, the practical approach to filter design is first to obtain the required magnitude characteristic for an application. If the delay characteristic is also an important consideration and the delay characteristic associated with the network function that gives the required magnitude characteristic is not satisfactory, the usual solution is to introduce additional delays so that the total delay is nearly flat over the frequency band in which a constant delay is important. These additional delays are furnished by the *delay equalizer* (or, equivalently, the *phase linearizer*). A delay equalizer usually consists of one or more allpass networks each of which has the same gain for all frequencies as shown in Fig. 1.2(e). They do not disturb the magnitude characteristic.

This procedure can be illustrated by an example. Suppose the lowpass magnitude characteristic is furnished by a normalized third-order Butterworth filter with the pass band located at $\omega_p = 0.9$. This corresponds to a passband ripple of 1.85 dB. The group delay of such a filter in the pass band is shown as curve 1 in Fig. 3.13. Since the variation of the delay is fairly smooth, it is estimated that the addition of a first-order allpass network function of the form

$$\frac{s-b}{s+b} \tag{3.37}$$

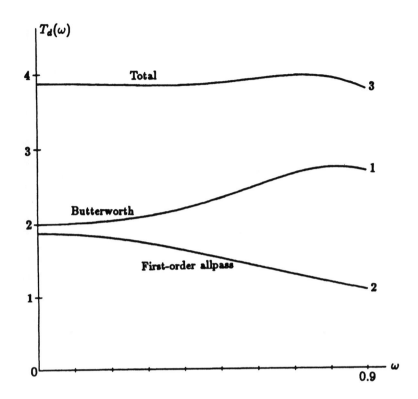

Figure 3.13: An example of delay equalization.

which has a delay function equal to

$$\frac{2b}{b^2 + \omega^2} \tag{3.38}$$

will be sufficient to equalize the delay. By trial and error, we find that $b = 1.065$ will be a good value to use. The additional delay introduced by this allpass function is shown as curve 2 in Fig. 3.13. The overall delay is shown as curve 3. It is seen that this total delay is quite flat.

If the delay to be equalized varies quite widely as some of the curves in Fig. 3.12 may suggest, it may be necessary to introduce several allpass functions, most of them of the general second-order form

$$\frac{s^2 - b_1 s + b_0}{s^2 + b_1 s + b_0} = \frac{s^2 - \dfrac{\omega_0}{Q}s + \omega_0^2}{s^2 + \dfrac{\omega_0}{Q}s + \omega_0^2} \tag{3.39}$$

in which $\omega_0 = \sqrt{b_0}$ and $Q = \sqrt{b_0}/b_1$. The second form of the expression in (3.39), in which the parameters are ω_0 and Q rather than b_1 and b_0, is preferable in indicating the variation of the associated delay functions. The parameter Q indicates the general shape of the delay functions. When Q is low (approximately 0.577 or lower), the delay is monotonically decreasing. When Q is high, the delay will have a peak. The higher Q is, the higher and sharper the peak will be. The parameter ω_0 gives the approximate location where the peak occurs. Figure 3.14 gives the delay

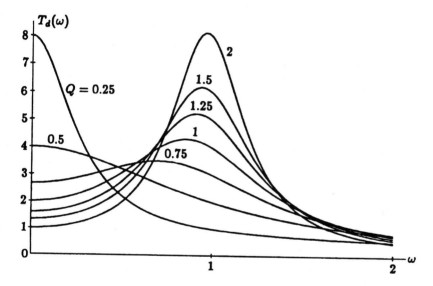

Figure 3.14: Delay characteristics of second-order allpass functions.

characteristics of second-order allpass functions for several values of Q for $\omega_0 = 1$.

By adding a number of second-order allpass functions with different Q's and ω_0's, it is possible to improve the delay characteristics substantially. However, when several functions are added, it would be difficult to rely on trial and error to choose their parameters. Usually a simple computer program incorporating certain optimization algorithms will be necessary to determine the number functions and their parameters or coefficients to achieve the desired equalization.

3.7   Summary

In this chapter, we first described a procedure by which a network function $H(s)$ can be obtained when its magnitude function $|H(j\omega)|$ is given. One key step is to factor a polynomial which may be of fairly high degree. The software MATLAB is suitable for this purpose. There are numerous other programs that can also be used.

Then we gave the closed-form formulas for poles of the Butterworth and Chebyshev lowpass filters. We also gave a qualitative description, using the conformal transformation, of how network functions for elliptic filters can be obtained. For this class of filters, we generally rely on tables or certain software for occasional needs. Several tables for all three types of filters are given in Appendix A.

Problems

3.1 Each of the following functions is the $|H(j\omega)|^2$ of a certain network function $H(s)$. Obtain all possible $H(s)$ for each given $|H(j\omega)|^2$.

(a) $\dfrac{\omega^4 - 5\omega^2 + 9}{\omega^4 + 10\omega^2 + 9}$

(b) $\dfrac{\omega^4 + 25}{\omega^4 + 12\omega^2 + 49}$

(c) $\dfrac{1}{(\omega^4 + 13\omega^2 + 36)(\omega^4 + 6\omega^2 + 25)}$

(d) $\dfrac{\omega^2 + 4}{(\omega^2 + 9)(\omega^4 + 21\omega^2 + 4)}$

(e) $\dfrac{\omega^4 - 2\omega^2 + 5}{\omega^{10} + 1}$

(f) $\dfrac{1}{\omega^6 + 7\omega^4 + 20\omega^2 + 20}$

(g) $\dfrac{1}{\omega^8 + 32\omega^6 - 58\omega^4 + 304\omega^2 + 9}$

3.2 Using only Equation (3.9), obtain the normalized Butterworth polynomial for a fourth-order lowpass filter in both the factored and the expanded forms. Verify your answer using MATLAB.

3.3 Using only Equation (3.9), obtain the normalized Butterworth polynomial for a sixth-order lowpass filter in both the factored and the expanded forms. Verify your answer using MATLAB.

3.4 Show that the Butterworth polynomials for even n are given by

$$Q(s) = \prod_{k=-1}^{-n/2} \left\{ s^2 - \left[2\cos\left(\frac{2k - n + 1}{2n}\pi \right) \right] s + 1 \right\}$$

What are the polynomials if n is odd?

3.5 The following network can be used to realize a third-order Butterworth lowpass filter with the network function

$$H(s) = \frac{E_2}{E_1} = \frac{K}{s^3 + 2s^2 + 2s + 1}$$

Derive the expression for $H(s)$ in terms of the circuit element values. Then, by matching the coefficients, solve for C_1, L_2, and C_3. Finally, evaluate K.

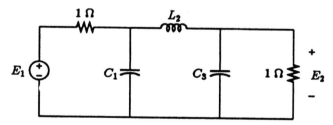

3.6 Using only Equation (3.18), determine the network function $H(s)$ such that it gives a Chebyshev lowpass characteristic with $\alpha_p = 0.4$ dB, $n = 3$, and $H(0) = 1$. Verify your answer using MATLAB.

3.7 Using only Equation (3.18), determine the network function $H(s)$ such that it gives a Chebyshev lowpass characteristic with $\alpha_p = 1.5$ dB, $n = 4$, and $|H(j\omega)|_{\max} = 1$. Verify your answer using MATLAB.

3.8 Using only Equation (3.18), determine the network function $H(s)$ such that it gives a Chebyshev lowpass characteristic with $\alpha_p = 1.2$ dB, $n = 6$, and $H(0) = 2$. Verify your answer using MATLAB.

3.9 Find the network function whose magnitude is given in Problem 2.20.

3.10 Find the network function whose magnitude is given in Problem 2.21.

3.11 Give the general form of the network function whose magnitude squared is defined in (2.31) with the $R_N(\omega)$ given in (2.29). Use as many known quantities as possible. What are the values of the network function at $s = 0$ and $s = \infty$?

3.12 Give the general form of the network function whose magnitude squared is defined in (2.31) with the $R_N(\omega)$ given in (2.30). Use as many known quantities as possible. What are the values of the network function at $s = 0$ and $s = \infty$?

3.13 Obtain (3.33) by the method of forcing the delay function to be maximally flat at the origin.

3.14 Obtain $y_5(s)$ by the method of truncating the continued fraction expansion of e^{-s}.

Chapter 4

Frequency transformation

Thus far, we have placed our major emphasis on the lowpass filters. One justification for this approach is that other types of filters can be generated from lowpass ones. We shall now develop these procedures to show how other types of filters can be obtained from lowpass ones. The procedures we are going to use are the *frequency transformations*. In developing each of these procedures, we will be addressing the following issues: (1) the effects on the magnitude characteristic, (2) the effects on the network function, (3) the effects on the poles and zeros, and (4) the effects on the network elements.

A frequency transformation amounts to rearranging the points on the j axis to achieve different filtering characteristics. As such, it is necessary for us to look at the entire j axis - both the positive and the negative halves. We shall use the complex variable $s = \sigma + j\omega$ for the normalized lowpass function. The transformed network functions will be functions of the variable $S = \Sigma + j\Omega$.

We must acknowledge that, although other types of filters can be obtained from lowpass ones by the frequency transformations outlined in this chapter, these procedures are not the only ways to obtain those other types of filters. For example, methods exist such that bandpass filters can be obtained directly. In fact, there are situations in which bandpass filters obtained by frequency transformation may not be practical and they must be obtained by other special methods. However, for many applications, the procedures presented here are quite satisfactory.

4.1   Lowpass-to-highpass transformation

We start with a normalized lowpass prototype filter whose pass band is $0 < \omega < \omega_p$ and $\omega_p = 1$ with a network function $H_{\mathrm{LP}}(s)$. We utilize the transform

$$s = \frac{\Omega_0}{S} \tag{4.1}$$

Along the imaginary axis

$$\omega = -\frac{\Omega_0}{\Omega} \tag{4.2}$$

With the substitution (4.1), we obtain a highpass network function

$$H_{\mathrm{HP}}(S) = H_{\mathrm{LP}}\left(\frac{\Omega_0}{S}\right)$$

To illustrate what this transform does to the magnitude characteristic, we start with the lowpass characteristic $|H_{\mathrm{LP}}(j\omega)|$ of Fig. 4.1. It is transformed into the highpass characteristic $|H_{\mathrm{HP}}(j\Omega)|$ of Fig. 4.1. The

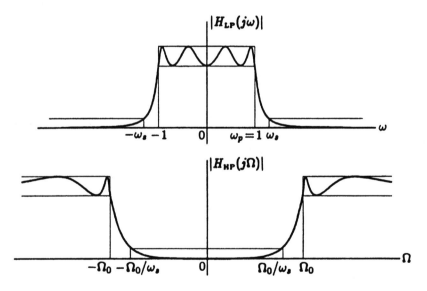

Figure 4.1: The lowpass-to-highpass transformation.

point $\omega = 1$ is mapped to $\Omega = -\Omega_0$ and the point $\omega = \omega_s$ to $\Omega = -\Omega_0/\omega_s$. Hence not only the lowpass characteristic is transformed into a highpass characteristic, but also there is a flip-over of the positive and negative halves of the characteristic. Because of the symmetry of the magnitude characteristic, this transform puts the pass band in the range $\Omega_0 < \Omega < \infty$ and the stop band in $0 < \Omega < \Omega_0/\omega_s$. Mathematically, the pass band is actually the mapping of $-1 < \omega < 0$, and the stop band that of $-\infty < \omega < -\omega_s$. As a practical matter, this fine detail is not important. However, it is as well to keep this mathematical detail in mind.

If $H_{\text{LP}}(s)$ has a pole (or a zero) at s_k, $H_{\text{HP}}(S)$ will have a pole (or zero) at

$$S_k = \frac{\Omega_0}{s_k} \qquad (4.3)$$

If the lowpass filter circuit has already been obtained, to obtain the highpass filter we can simply modify the lowpass filter by making the following observation. All resistances are unchanged since they are not frequency-sensitive. For an inductance of L_i henrys, it has an impedance of sL_i ohms. This impedance becomes $\Omega_0 L_i/S$ ohms in the highpass filter and corresponds to the impedance of a capacitance of $1/\Omega_0 L_i$ farads. Analogously, a capacitance of C_i farads in the lowpass filter has an admittance of sC_i siemens. It becomes an admittance of $\Omega_0 C_i/S$ siemens in the highpass filter and corresponds to the admittance of an inductance of $1/\Omega_0 C_i$ henrys. These element transformations are summarized in Table 4.1.

EXAMPLE 4.1. Obtain the transfer function of the fourth-order Butterworth highpass filter with $H_{\text{HP}}(\infty) = 1$ and $\Omega_0 = 2\pi \times 10^4$ rad/sec.

SOLUTION From Table A.2, the normalized lowpass transfer function is

$$H_{\text{LP}}(s) = \frac{1}{s^4 + 2.61313s^3 + 3.41421s^2 + 2.61313s + 1}$$

Replacing every s with $2\pi \times 10^4/S$, we obtain

$$H_{\text{HP}}(S) = 1 \left/ \left[\left(\frac{2\pi \times 10^4}{S}\right)^4 + 2.61313\left(\frac{2\pi \times 10^4}{S}\right)^3 \right.\right.$$

$$\left.\left. + 3.41421\left(\frac{2\pi \times 10^4}{S}\right)^2 + 2.61313\left(\frac{2\pi \times 10^4}{S}\right) + 1 \right]$$

$$= S^4/[S^4 + 1.64185 \times 10^5 S^3 + 1.34787 \times 10^{10} S^2$$

$$+ 6.48182 \times 10^{14} S + 1.55854 \times 10^{19}]$$

The lowpass-to-highpass transform can also be effected by using the MATLAB command lp2hp, which has the format

$$[\text{numt}, \text{dent}] = \text{lp2hp}(\text{num}, \text{den}, \text{Wo})$$

where num is the vector containing the numerator coefficients and den is that containing the denominator coefficients of the lowpass prototype network function with $\omega_p = 1$. Wo is the cutoff frequency (or Ω_0) of the highpass characteristic. The command returns numerator and denominator coefficient vectors, numt and dent respectively. The steps corresponding to the above solution using MATLAB will be

```
% Specify numerator and denominator coefficient vectors
> num = 1;
> den = [1 2.61311 3.41420 2.61311 1];
> format short e;
% Specify cutoff frequency
> Wo = 2*pi*1.0e4;
> [numt,dent] = lp2hp(num,den,Wo)
numt = 1.0000e+000  0  0  0  0
% Coefficients of lower-degree terms of numt should be
% and are artificially set to zero
dent = 1.0000e+000  1.6419e+005  1.3479e+010  6.4818e+014
       1.5585e+019
```

4.2   Lowpass-to-bandpass transformation

Again, we start off with a normalized ($\omega_p = 1$) lowpass prototype filter with network function $H_{\text{LP}}(s)$. To obtain a bandpass filter, we use the transformation

$$s = \frac{S^2 + \Omega_0^2}{BS} \qquad (4.4)$$

The bandpass filter function is

$$H_{\text{BP}}(S) = H_{\text{LP}}\left(\frac{S^2 + \Omega_0^2}{BS}\right) \qquad (4.5)$$

Along the imaginary axes

$$j\omega = \frac{-\Omega^2 + \Omega_0^2}{Bj\Omega}$$

or

$$\Omega^2 - B\omega\Omega - \Omega_0^2 = 0 \tag{4.6}$$

Thus the point $\omega = 0$ maps onto $\Omega = \pm\Omega_0$. The points $\omega = \pm\infty$ map onto $\Omega = 0$ and $\Omega = \pm\infty$. For $\omega = \omega_p = 1$, we solve the quadratic equation

$$\Omega^2 - B\Omega - \Omega_0^2 = 0$$

which gives

$$\Omega = \frac{B}{2} \pm \sqrt{\frac{B^2}{4} + \Omega_0^2} \tag{4.7}$$

One of the values of Ω in (4.7) is positive and one negative. We let

$$\Omega_2 = \frac{B}{2} + \sqrt{\frac{B^2}{4} + \Omega_0^2} \tag{4.8}$$

$$-\Omega_1 = \frac{B}{2} - \sqrt{\frac{B^2}{4} + \Omega_0^2} \tag{4.9}$$

If we let $\omega = -1$ in (4.6), we would obtain two roots equal to $-\Omega_2$ and Ω_1. Hence a lowpass band is mapped onto two bandpass regions, one on the positive axis and one on the negative axis. This mapping is shown in Fig. 4.2. From (4.8) and (4.9), we also see that

$$B = \Omega_2 - \Omega_1$$

which is the *bandwidth* of the bandpass filter. Equations (4.8) and (4.9) also give

$$\Omega_1\Omega_2 = \left[\frac{B}{2} + \sqrt{\frac{B^2}{4} + \Omega_0^2}\right]\left[-\frac{B}{2} + \sqrt{\frac{B^2}{4} + \Omega_0^2}\right] = \Omega_0^2$$

or

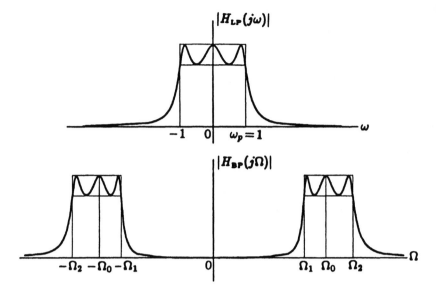

Figure 4.2: The lowpass-to-bandpass transformation.

$$\frac{\Omega_0}{\Omega_1} = \frac{\Omega_2}{\Omega_0} \tag{4.10}$$

Hence the upper band edge, Ω_2, and the lower band edge, Ω_1, are geometrically symmetric about the center frequency, Ω_0. In fact it can be shown that the entire bandpass characteristic is geometrically symmetric about Ω_0. Thus if the bandpass characteristic is plotted versus $\log \Omega$, then the entire curve will be symmetric about $\log \Omega_0$. The frequency Ω_0 is known as the *band center* or *center frequency* of the bandpass characteristic. The ratio of the band center to the bandwidth is known as the *quality factor*, Q, of the bandpass filter. The higher Q is, the narrower the bandwidth of a bandpass filter will be.[1]

$$Q = \frac{\Omega_0}{B}$$

If the lowpass filter function has a pole at s_k, then this transform maps it onto two poles. We let

---

[1]The same letter will be used later in active filters to denote a somewhat related parameter in a quadratic polynomial - the pole or zero Q. The significance of Q in that application is quantitatively slightly different from the bandpass filter Q. Whether Q denotes a bandpass Q or a pole or zero Q is usually clear from the context. We shall not use different notation for these two similar but different definitions.

$$s_k = \frac{S^2 + \Omega_0^2}{BS}$$

which may be rewritten as

$$S^2 - s_k BS + \Omega_0^2 = 0 \tag{4.11}$$

Solution of (4.11) gives the two corresponding bandpass filter function poles in the S plane.

In using this transformation, the bandpass filter can be obtained from its lowpass prototype by replacing the energy-storing elements of the latter. For an inductance of L_i henrys in the prototype, its impedance becomes $\frac{S^2+\Omega_0^2}{BS} L_i = \frac{L_i}{B} S + \frac{\Omega_0^2 L_i}{BS}$ ohms, which can be identified as an inductance of L_i/B henrys in series with a capacitance of $B/\Omega_0^2 L_i$ farads. For a capacitance of C_i farads, its admittance becomes $\frac{S^2+\Omega_0^2}{BS} C_i = \frac{C_i}{B} S + \frac{\Omega_0^2 C_i}{BS}$ siemens, which can be identified as a capacitance of C_i/B farads in parallel with an inductance of $B/\Omega_0^2 C_i$ henrys. These replacements are summarized in Table 4.1.

EXAMPLE 4.2. The filter shown in Fig. 4.3 is a fourth-order Chebyshev lowpass filter with $\alpha_p = 1$ dB, and $\omega_p = 1$. Obtain from that filter, a bandpass filter with $\Omega_0 = 400$ and $B = 150$. Determine Ω_2 and Ω_1. Obtain the bandpass filter, its voltage transfer function and its poles.

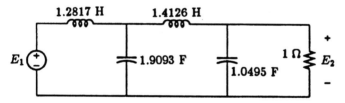

Figure 4.3: A fourth-order Chebyshev lowpass filter.

SOLUTION We use the transform

$$s = \frac{S^2 + 400^2}{150S}$$

From (4.8) and (4.9),

$$\Omega_2 = \frac{150}{2} + \sqrt{\frac{150^2}{4} + 400^2} = 481.97$$

$$\Omega_1 = -\frac{150}{2} + \sqrt{\frac{150^2}{4} + 400^2} = 331.97$$

From Table 4.1, we obtain the bandpass filter of Fig. 4.4.

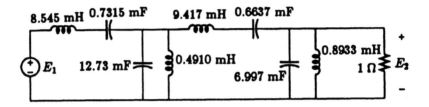

Figure 4.4: The bandpass filter obtained from the lowpass filter of Fig. 4.3.

From Table A.8, the network function of the lowpass filter is

$$H_{\mathrm{LP}}(s) = \frac{E_2}{E_1} = \frac{K}{s^4 + 0.952811s^3 + 1.453925s^2 + 0.742619s + 0.275628}$$

Because $E_2/E_1 = 1$ at $s = 0$, $K = 0.275628$. The network function for the bandpass filter is

$$H_{\mathrm{BP}}(S) = K \left/ \left[\left(\frac{S^2+400^2}{150S}\right)^4 + 0.952811\left(\frac{S^2+400^2}{150S}\right)^3 \right.\right.$$

$$\left. + 1.453925\left(\frac{S^2+400^2}{150S}\right)^2 + 0.742619\left(\frac{S^2+400^2}{150S}\right) + 0.275628 \right]$$

$$= 150^4 \times 0.275628S^4 / [S^8 + 142.921S^7 + 6.72712 \times 10^5 S^6$$

$$+ 7.11087 \times 10^7 S^5 + 1.64207 \times 10^{11} S^4 + 1.13773 \times 10^{13} S^3$$

$$+ 1.72214 \times 10^{16} S^2 + 5.85407 \times 10^{17} S + 6.5536 \times 10^{20}]$$

From Table A.7, the poles of the lowpass filter function are located at $-0.139536 \pm j0.983379$ and $-0.336870 \pm j0.407329$. From (4.11), two of the poles of the bandpass function can be obtained by solving the equation

$$S^2 - 150(-0.139536 + j0.983379)S + 400^2 = 0$$

which gives $-12.3634 + j480.366$ and $-8.56697 - j332.859$. Two other poles can be obtained by solving the quadratic

$$S^2 - 150(-0.336870 - j0.407329)S + 400^2 = 0$$

which gives $-23.3374 + j369.823$ and $-27.1931 - j430.923$. Since the poles of $H_{\text{BP}}(S)$ must occur in conjugate pairs, it is not necessary to obtain the other four poles. They are simply the conjugates of the four poles obtain above.

It is also possible to reconstruct the bandpass transfer function by using the eight poles just calculated. The numerator will be of the form $K'S^4$ and K' must be adjusted so that the gain at $S = j400$ is equal to unity.

An alternative to the above procedure would be first to obtain a normalized bandpass filter ($\Omega_0 = 1$) and its associated functions and poles. Using this approach, we would first apply the transformation

$$s = \frac{S^2 + 1}{0.375S}$$

After all quantities have been obtained, we can then apply a frequency-scaling factor $k_f = 400$.

The lowpass-to-bandpass transform can also be effected by using MAT-LAB command lp2bp, which has the format

$$[\text{numt,dent}] = \text{lp2bp(num,den,Wo,Bw)}$$

where num and den are the numerator and denominator coefficient vectors of the lowpass prototype network function with $\omega_p = 1$. Wo is the band-center frequency (or Ω_0) and Bw is the bandwidth (or B) of the bandpass characteristic. The command returns numerator and denominator coefficient vectors, numt and dent, respectively. The steps corresponding to the above solution using MATLAB will be

```
>> num = 0.275628;
>> den = [1 0.952811 1.453925 0.742619 0.275628];
>> format short e;
>> Wo = 400;
>> Bw = 150;
>> [numt,dent] = lp2bp(num,den,Wo,Bw)
numt = 0 0 0 0 1.3954e+008 0 0 0 0
% Coefficients other than the s^4 term should be zero
```

```
% and are artificially set to zero
dent = 1 1.4292e+002 6.7271e+005 7.1109e+007 1.6421e+011
        1.1377e+013 1.7221e+016 5.8541e+017 6.5536e+020
```

4.3   Lowpass-to-bandreject transformation

To obtain a bandreject filter from a lowpass prototype, we use the transformation

$$s = \frac{BS}{S^2 + \Omega_0^2}$$ (4.12)

Along the j axis

$$j\omega = \frac{Bj\Omega}{-\Omega^2 + \Omega_0^2}$$

or

$$\Omega^2 + \frac{B}{\omega}\Omega - \Omega_0^2 = 0$$ (4.13)

Since in a bandreject application we are usually concerned about the frequency band in which signals are rejected, our focus should be on what happens to the stop band. Hence it is most expedient that we first normalize the lowpass prototype filter such that its stop band is $\omega > \omega_s = 1$, as shown in Fig. 4.5. The points $\omega = \pm 1$ are mapped to the $j\Omega$ axis by solving the quadratic equations

$$\Omega^2 \pm B\Omega - \Omega_0^2 = 0$$

The solution of these two equations will render four values of Ω which, if taken as a group, are identical to those given in (4.8) and (4.9) and their negatives. Figure 4.5 illustrates this transformation.

If the lowpass filter function has a pole (or a zero) at s_k, the locations of the corresponding poles (or zeros) of the bandreject filter function can be found by solving the equation

$$S^2 - \frac{B}{s_k}S + \Omega_0^2 = 0$$

To obtain the transfer function of a bandreject filter from that of a lowpass prototype, we can also use the MATLAB command lp2bs. The

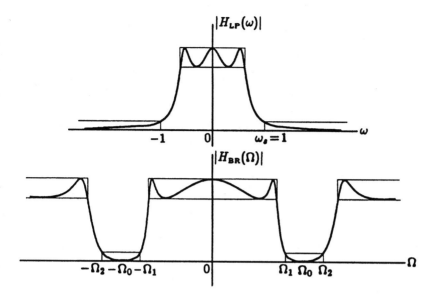

Figure 4.5: The lowpass-to-bandreject transformation.

Table 4.1. Element replacements in frequency transformations

Normalized lowpass filter elements	Highpass filter elements	Bandpass filter branches	Bandreject filter branches
L_i	$\dfrac{1}{\Omega_0 L_i}$	$\dfrac{L_i}{B} \quad \dfrac{B}{\Omega_0^2 L_i}$	$\dfrac{BL_i}{\Omega_0^2}$ $\dfrac{1}{BL_i}$
C_i	$\dfrac{1}{\Omega_0 C_i}$	$\dfrac{B}{\Omega_0^2 C_i}$ $\dfrac{C_i}{B}$	$\dfrac{1}{BC_i} \quad \dfrac{BC_i}{\Omega_0^2}$

use of this command is similar to the lowpass-to-bandpass transform command lp2bp.

The bandreject filter can be obtained by replacing the energy-storing elements in the lowpass prototype. An inductance of L_i henrys will have an impedance of $\frac{BS}{S^2+\Omega_0^2}L_i$ ohms or an admittance of $\frac{S}{BL_i} + \frac{\Omega_0^2}{BL_iS}$ siemens, which can be identified as the parallel combination of a capacitance of $1/BL_i$ farads and an inductance of BL_i/Ω_0^2 henrys. Analogously, a capacitance of C_i farads becomes the series combination of an inductance of $1/BC_i$ henrys and a capacitance of BC_i/Ω_0^2 farads. These replacements, as well as replacements mentioned earlier for other transformations, are summarized in Table 4.1.

4.4 Summary

In this chapter, we have outlined three transformations that will enable us to generate highpass, bandpass, and bandreject filters from a lowpass prototype. The transformations as well as some mathematical details are delineated. It should be kept in mind that these transformations are merely the most expeditious methods of obtaining those other types of filters from lowpass ones. For most everyday uses, filters obtained this way are often adequate. However, these methods do have their limitations and drawbacks. Occasions do arise in specialized needs where the transform methods are not suitable or satisfactory. Certain applications such as channel-separation bandpass filters will require specialized design techniques. These and other specialized techniques are beyond the scope of this text.

Problems

4.1 The lowpass filter below has a voltage transfer function, E_2/E_1, with a passband frequency $\omega_p = 1$ rad/sec. Transform it into a highpass one such that $\Omega_p = 10,000$ rad/sec. Then impedance-scale it so the capacitance values become 1 μF and 3μF.

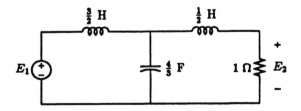

4.2 Locate the poles of the voltage transfer function of the highpass filter obtained in Problem 4.1.

4.3 Obtain a bandpass filter from the lowpass filter of Problem 4.1 so that the band center is located at $\Omega_0 = 1,000$ rad/sec and the bandwidth is $B = 500$ rad/sec.

4.4 Locate the poles of the voltage transfer function of the bandpass filter obtained in Problem 4.3.

4.5 The following circuit is a Chebyshev lowpass filter with $\omega_p = 1$ and $\alpha_p = 0.01$ dB. Obtain a bandpass filter with the same α_p and a pass band between 50 and 100 Hz to work between two resistances of 50 Ω and 100 Ω.

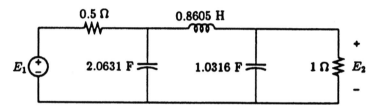

4.6 A lowpass filter with the magnitude squared function

$$|H_{\text{LP}}(s)|^2 = \frac{1}{1 + \omega^6}$$

has already been designed. This lowpass filter is transformed into a bandpass one by using the transform

$$s = \frac{S^2 + 100^2}{10S}$$

Where will the value of $|H_{\text{BP}}(S)|^2$ be equal to 0.1 along the Ω axis?

4.7 A bandpass filter has been obtained from a lowpass prototype by the transformation

$$s = \frac{S^2 + 100^2}{5S}$$

It is known that the bandpass filter function has a pair of poles at $-1 \pm j102$. Determine another pair of poles of the bandpass filter function.

4.8 The network function

$$H(s) = \frac{s^3}{s^3 + 100s^2 + 3 \times 10^4 s + 10^6}$$

is that of a highpass filter with its pass band located in the range $\omega \geq 100$ rad/sec. It is desired to obtain a band pass filter from this highpass one with the same passband ripple. The band center is to be at 500 rad/sec and the bandwidth 125 rad/sec. Describe the transformation steps to accomplish this and obtain the transfer function of the bandpass filter.

4.9 A certain lowpass filter has already been designed with its magnitude characteristic $|H_{LP}(j\omega)|$ shown below. The transform

$$s = \frac{S^2 + 269}{6S}$$

is applied to this function to obtain a bandpass characteristic. Give a sketch of the bandpass magnitude characteristic. Give all crucial points on this characteristic.

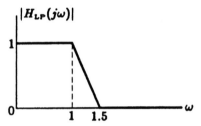

4.10 Repeat Problem 4.9 if the transform is to be the lowpass-to-bandreject one. Namely

$$s = \frac{6S}{S^2 + 269}$$

4.11 A Butterworth lowpass filter with the following magnitude squared characteristic

$$|H_{LP}(j\omega)|^2 = \frac{1}{1 + \omega^8}$$

has already been realized. This lowpass filter is transformed into a bandreject one by the transform

$$s = \frac{7S}{S^2 + 75^2}$$

At what values of Ω will the gain of the bandreject filter be equal to 0.5?

4.12 Consider the lowpass filter in Problem 4.1 to have a stop band in the range $2.5 < w < \infty$. Obtain a bandreject filter such that its stop band is in the range $100 < \Omega < 500$.

4.13 A fourth-order Butterworth filter has the magnitude squared

$$|H_{LP}(j\omega)|^2 = \frac{1}{1 + \omega^8}$$

It is to be transformed into a bandreject filter by the transform

$$s = \frac{500S}{S^2 + 10^7}$$

Where are the poles and zeros of the new filter function in the S plane?

4.14 The magnitude of a certain filter function $H(s)$ is given in the graph below. Give a dimensioned sketch of the magnitude characteristic $|H'(j\Omega)|$, where

$$H'(S) = H\left(\frac{S^2 + 20}{5S}\right)$$

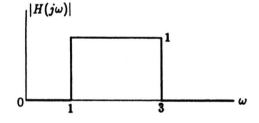

Chapter 5

Properties and synthesis of passive networks

We are now at a point where we can assume that a network function $H(s)$ or, in some cases, its magnitude function $|H(\omega)|$ has been determined. Our next task is to find one or more networks that will realize the given function. We will cover two broad categories of networks. One of them is networks using lossless elements. The other is networks containing resistors, capacitors, and op amps.

Before we embark on the development of various techniques to realize filters, we shall state without proof certain properties of some lossless networks and synthesis techniques that are relevant to filter realizations using lossless elements.

5.1 The driving-point function of a passive oneport - the positive real function

It can be shown that the driving-point impedance or admittance of a passive network (networks containing resistors, inductors, capacitors, mutual inductances, and ideal transformers)[1] must be *positive real*. In addition to being a real rational function, a function $F(s)$ is positive real if and only if [Gu]

---

[1] Other devices such as the gyrator and the isolator are also passive and can be included here.

(1) $F(s)$ is real if s is real.

(2) $\text{Re}[F(s)] \geq 0$ if $\text{Re}[s] \geq 0$.

The following necessary conditions are implied by the positive-real property:

(1) The reciprocal of a positive-real function is another positive-real function.

(2) $F(s)$ can have no right-halfplane poles or zeros.

(3) $\text{Re}[F(j\omega)] \geq 0$.

(4) If $F(s)$ has a pole on the j axis, it must be simple with a real and positive residue. [The residue in a pole at s_i is the coefficient k_i in the term $k_i/(s - s_i)$ of the partial-fraction expansion of $F(s)$].

The study of positive-real functions was an important milestone in the evolution of the area of passive network synthesis. For the realization of filters, the positive-real functions themselves are not important and we will not elaborate on them.

5.2   The driving-point function of a lossless oneport - the lossless function

When a network contains only lossless elements, its driving-point function is a limiting case of the positive-real function - it is a positive-real function for which

$$\text{Re}[F(j\omega)] \equiv 0 \tag{5.1}$$

We shall call these functions *lossless functions*. This term should be thought of as a contraction of *the driving-point function of a lossless network*.

5.2.1   Properties of a lossless function

Each of the following is a necessary condition for a real rational function

$$F(s) = \frac{P(s)}{Q(s)} \tag{5.2}$$

to be a lossless function [Gu].

(1) It is the ratio of an even polynomial to an odd polynomial, or vice versa.

(2) All poles and zeros lie on the j axis, they are simple, and they alternate.[2]

(3) It must have either a pole or a zero at $s = 0$. Also at $s = \infty$.

(4) $|\deg[P(s)] - \deg[Q(s)]| \equiv 1$.

(5) $P(s)$ and $Q(s)$ have either all even-powered terms or all odd-powered terms, and no intermediate terms may be missing.

(6) The reciprocal of a lossless function is another lossless function.

(7) $P(s) + Q(s)$ is a Hurwitz polynomial - all its zeros lie in the left halfplane.

(8) On the j axis, $F(j\omega) = jX(\omega)$ is purely imaginary. The function $X(\omega)$ is either a reactance function or a susceptance function.[3] It can be shown that

$$\frac{dX(\omega)}{d\omega} > 0 \tag{5.3}$$

In conjunction with Property (1), Property (2), (7), or (8) also constitutes the sufficient conditions for a function to be a lossless function.

Depending on whether a lossless function has a pole of a zero at either the origin or infinity, a lossless function may have one of the forms of the examples given in Table 5.1.

Sometimes it is useful to give a sketch of the imaginary part of the lossless function, $X(\omega)$, along the $j\omega$ axis. For each of the four types of functions in Table 5.1, the variation of its imaginary part is shown in Fig. 5.1.

---

[2]Their residues must be real and positive because of the positive real conditions mentioned in the previous section.

[3]Because of Property (6), we could deal with either an impedance function or an admittance function. Arbitrarily, we choose to deal mostly with the impedance function.

Table 5.1: Examples of lossless functions

Type	Lossless functions	At $s = 0$	At $s = \infty$
(a)	$\dfrac{(s^2 + 1)(s^2 + 3)(s^2 + 5)}{s(s^2 + 2)(s^2 + 4)}$	pole	pole
(b)	$\dfrac{(s^2 + 2)(s^2 + 5)}{s(s^2 + 3)(s^2 + 7)}$	pole	zero
(c)	$\dfrac{s(s^2 + 6)(s^2 + 20)}{(s^2 + 3)(s^2 + 9)}$	zero	pole
(d)	$\dfrac{s(s^2 + 8)(s^2 + 20)}{(s^2 + 4)(s^2 + 10)(s^2 + 30)}$	zero	zero

5.2.2 Foster's expansion of a lossless function

A lossless function written in partial-fraction form will be

$$F(s) = \frac{k_0}{s} + k_\infty s + \sum_i \left[\frac{c_i}{s + j\omega_i} + \frac{c_i}{s - j\omega_i} \right] \tag{5.4}$$

The term k_0/s represents the pole of $F(s)$ at the origin. If $F(s)$ has no pole at the origin, then $k_0 = 0$. Similarly, $k_\infty s$ represents the pole, if there is one, at infinity. Each term in the summation in (5.4) represents a pair of finite nonzero poles. There residues are real and positive, and they are equal. It is more convenient to combine each pair of such terms. Equation (5.4) will then read

$$F(s) = \frac{k_0}{s} + k_\infty s + \sum_i \frac{k_i s}{s^2 + \omega_i^2} \tag{5.5}$$

where $k_i = 2c_i$. Expression (5.5) is known as *Foster's expansion* of a lossless function. It is clear that

$$k_0 = sF(s)|_{s=0} \tag{5.6}$$

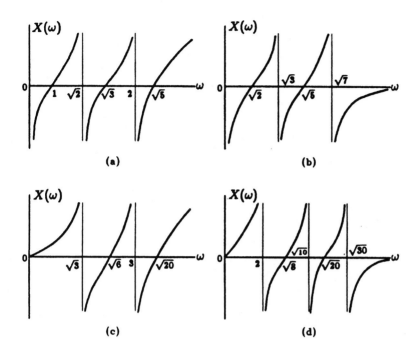

Figure 5.1: Plots of the imaginary parts of the four types of lossless functions given in Table 5.1.

$$k_\infty = \left. \frac{F(s)}{s} \right|_{s=\infty} \tag{5.7}$$

To obtain k_i we multiply (5.5) by $(s^2 + \omega_i^2)/s$ and then set $s^2 = -\omega_i^2$ to get

$$k_i = \left[\frac{(s^2 + \omega_i^2)}{s} F(s) \right]_{s^2 = -\omega_i^2} \tag{5.8}$$

EXAMPLE 5.1. Obtain the Foster's expansion of

$$F(s) = \frac{(s^2 + 1)(s^2 + 5)(s^2 + 20)}{s(s^2 + 2)(s^2 + 10)} = \frac{s^6 + 26s^4 + 125s^2 + 100}{s^5 + 12s^3 + 20s} \tag{5.9}$$

SOLUTION

$$k_0 = \frac{1 \times 5 \times 20}{2 \times 10} = 5$$

$$k_\infty = 1$$

$$k_1 = \left. \frac{(s^2 + 1)(s^2 + 5)(s^2 + 20)}{s^2(s^2 + 10)} \right|_{s^2 = -2} = \frac{(-1)(3)(18)}{(-2)(8)} = \frac{27}{8}$$

$$k_2 = \left. \frac{(s^2 + 1)(s^2 + 5)(s^2 + 20)}{s^2(s^2 + 2)} \right|_{s^2 = -10} = \frac{(-9)(-5)(10)}{(-10)(-8)} = \frac{45}{8}$$

Hence

$$F(s) = \frac{5}{s} + s + \frac{\frac{27}{8}s}{s^2 + 2} + \frac{\frac{45}{8}s}{s^2 + 10} \tag{5.10}$$

Foster's expansion of a lossless function can also be obtained by first obtaining the partial fraction of the function using MATLAB command

$$[\mathbf{r}, \mathbf{p}, \mathbf{k}] = \mathtt{residue(a,b)}$$

where a is the numerator coefficient vector and b is the denominator coefficient vector. This command returns the residues as a column vector r, the poles as a column vector p, and the coefficients of the quotient polynomial as a row vector k. For the above example, we have

```
% Specify numerator coefficient vector
» a = [1 0 26 0 125 0 100];
% Specify denominator coefficient vector
» b = [1 0 12 0 20 0];
» [r,p,k] = residue(a,b)
r = 2.8125, 2.8125, 1.6875, 1.6875, 5
p = 3.1623i, -3.1623i, 1.4142i, -1.4142i, 0
k = 1 0
```

Thus

$$F(s) = \frac{2.8125}{s - j3.1623} + \frac{2.8125}{s + j3.1623} + \frac{1.6875}{s - j1.4142} + \frac{1.6875}{s + j1.4142} + \frac{5}{s} + s + 0$$

which results in the same expansion as given in (5.10).

5.2.3 Foster's realizations of a lossless function

Foster's expansion of a lossless function not only reveals the content
in each pole (how strong each pole is, so to speak), it also enables us
to obtain inductance-capacitance (LC) networks whose impedance or
admittance is equal to a given lossless function. This is our first exposure
to the synthesis procedure. For example, suppose we wish to realize the
lossless function of (5.9) as an impedance. From (5.10), the first term
is an impedance of $5/s$ ohms, which is the impedance of a capacitor of
$1/5$ farad. The second term is an impedance of s ohms, which is the
impedance of an inductance of 1 henry.

The third term corresponds to a circuit of

$$\frac{\frac{27}{8}s}{s^2+2}\ \Omega \quad \text{or} \quad \frac{8}{27}s + \frac{16}{27s}\ \mho$$

which may be identified with the parallel combination of an $\frac{8}{27}$-F capacitor and a $\frac{27}{16}$-H inductor.

Similarly, the fourth term is

$$\frac{\frac{45}{8}s}{s^2+10}\ \Omega \quad \text{or} \quad \frac{8}{45}s + \frac{16}{9s}\ \mho$$

which corresponds to the parallel combination of an $\frac{8}{45}$-F capacitor and
a $\frac{9}{16}$-H inductor.

In summary, the circuit of Fig. 5.2 realizes an impedance function that
is equal to $F(s)$ of (5.9).

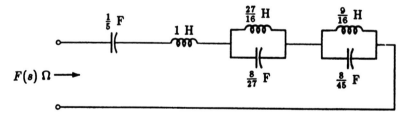

Figure 5.2: A realization of the impedance equal to the lossless function
of (5.9).

Alternatively, we may take the reciprocals of $F(s)$ and realize it as an
admittance. We have

$$Y(s) = \frac{1}{F(s)} = \frac{\frac{9}{76}s}{s^2+1} + \frac{\frac{1}{4}s}{s^2+5} + \frac{\frac{12}{19}s}{s^2+20} \ \mho \qquad (5.11)$$

Each term in (5.11) can be identified as a series LC branch. For instance, the first term is equal to

$$\frac{\frac{9}{76}s}{s^2+1} \ \mho \qquad \text{or} \qquad \frac{76}{9}s + \frac{76}{9s} \ \Omega$$

which is a $\frac{76}{9}$-henry inductor in series with a $\frac{9}{76}$-farad capacitor. Similar interpretation of the other two terms and the parallel combination of the three branches leads to the network of Fig. 5.3.

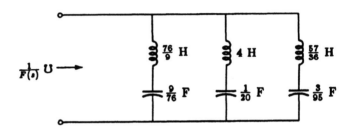

Figure 5.3: Another realization of the lossless function of (5.9).

It should be pointed out that although the network in Fig. 5.2 is synthesized on the basis of impedance while that of Fig. 5.3 on the basis of admittance, since the impedance of one is the reciprocal of the admittance of the other, their driving-point functions are identical. These two circuits are equivalent to each other as far as their driving-point functions are concerned. We shall refer to the realization of the type of Fig. 5.2 as *Foster 1* realization, and the realization of the type of Fig. 5.3 as the *Foster 2* realization. Foster 1 realization renders a network consisting of a number of branches in series. Foster 2 realization renders a network that is the parallel combination of a number of branches.

5.2.4   Removal of poles at infinity

If a lossless function $F(s)$ has a pole at infinity, the pole can be removed by subtraction. The remainder is

$$F_1(s) = F(s) - k_\infty s$$

The order of $F_1(s)$ is one lower than that of $F(s)$. $F_1(s)$ is another lossless function that does not have a pole at infinity.

For example, if

$$Z(s) = \frac{(s^2 + 4)(s^2 + 16)}{s(s^2 + 9)} \ \Omega \tag{5.12}$$

then

$$Z_1(s) = Z(s) - s = \frac{11s^2 + 64}{s(s^2 + 9)} \ \Omega \tag{5.13}$$

Since $Z_1(s)$ has no pole at infinity, it must have a zero there. Hence $Y_1(s) = 1/Z_1(s)$ has a pole there. We can remove the pole there again.

$$Y_2(s) = Y_1(s) - \frac{1}{11}s = \frac{\frac{35}{121}s}{s^2 + \frac{64}{11}} \ \mho$$

We can again remove the pole at infinity from $1/Y_2(s)$ to get

$$Z_3(s) = \frac{1}{Y_2(s)} - \frac{121}{35}s = \frac{704}{35s} \ \Omega$$

These pole removal steps can be carried out using MATLAB as follows.

```
» Z='(s^4 + 20*s^2 + 64)/(s^3 + 9*s)';
% Remove the pole at infinity (= s) from Z
» Z1=simplify(symsub(Z,'s'))
Z1 = (11*s^2 + 64)/s/(s^2 + 9)
% Let Y1=1/Z1
» Y1=simplify(symdiv(1,Z1))
Y1 = 1/(11*s^2 + 64)*s*(s^2 + 9)
```

Application of command **pretty(Y1)** yields

$$\frac{s(s^2 + 9)}{11s^2 + 64}$$

```
» Remove the pole at infinity (= s/11) from Y1
» Y2=simplify(symsub(y1,'11/s'))
Y2 = 35/11*s/(11*s^2 + 64)
```

Application of command **pretty(Y2)** yields

$$\frac{35}{11}\frac{s}{11s^2 + 64}$$

% Let Z2=1/Y2
> Z2=simplify(symdiv(1,Y2))
Z2 = 11/35*(11*s^2 + 64)/s

Application of command **pretty(Z2)** yields

$$\frac{11}{35}\frac{11s^2 + 64}{s}$$

% Remove the pole at infinity (= 121/35s) from Z2
> Z3=simplify(symsub(Z2,'121*s/35'))
Z3 = 704/35/s

Application of command **pretty(Z3)** yields

$$\frac{704}{35s}$$

For the purpose of synthesis, each pole removal can be identified as the extraction of a network element. A network is found when sufficient removal steps have been effected so that the remainder is either a single term or simple enough that it can easily be identified. In the example above, first an s-ohm impedance or a 1-henry series inductor was removed. Then a $\frac{1}{11}s$-mho admittance or a $\frac{1}{11}$-farad parallel capacitor was removed. And so on. Summarizing the three removal steps and the remainder, we obtain the network of Fig. 5.4.

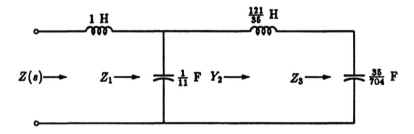

Figure 5.4: Network obtained by removing poles at infinity.

It can be observed that each removal of a pole at infinity is really the reduction of an improper fraction into a proper fraction. For example, (5.12) may be written as

$$\frac{s^4 + 20s^2 + 64}{s^3 + 9s} = s + \frac{11s^2 + 64}{s^3 + 9s}$$

which can also be accomplished by long division. The entire sequence of steps can be duplicated by continued long division as shown below.[4]

$$
\begin{array}{l}
\qquad\qquad s\ \Omega \\
s^3 + 9s\ \overline{\big)\ s^4 + 20s^2 + 64} \\
\qquad\qquad s^4 + 9s^2 \qquad\qquad \frac{1}{11}s\ \mho \\
\qquad\qquad \overline{\quad 11s^2 + 64\ }\ \big)\ s^3 + 9s \\
\qquad\qquad\qquad\qquad s^3 + \frac{64}{11}s \qquad \frac{121}{35}s\ \Omega \\
\qquad\qquad\qquad\qquad \overline{\quad \frac{35}{11}s\ }\ \big)\ 11s^2 + 64 \\
\qquad\qquad\qquad\qquad\qquad 11s^2 \qquad\qquad \frac{35}{704}s\ \mho \\
\qquad\qquad\qquad\qquad\qquad\qquad \overline{\quad 64\ }\ \big)\ \frac{35}{11}s \\
\qquad\qquad\qquad\qquad\qquad\qquad\qquad\qquad \frac{35}{11}s
\end{array}
$$

Alternatively, we may use a continued fraction to summarize the above results, namely

$$F(s) = s + \cfrac{1}{\frac{1}{11}s + \cfrac{1}{\frac{121}{35}s + \cfrac{1}{\frac{35}{704}s}}} \tag{5.14}$$

The realization by continued removal of poles at infinity is known as the *Cauer 1* realization.

The long division of two polynomials can be performed using the MAT-LAB command deconv which has the format

$$[q,r] = deconv(b,a)$$

in which b is coefficient vector of the dividend (numerator) polynomial and a that of the divisor (denominator) polynomial. The command returns the coefficient vector of the quotient polynomial q and the coefficient vector of the remainder polynomial r. The series of long divisions

---

[4]This algorithm is also used as a test of whether $P(s) + Q(s)$ is Hurwitz or not. For $P(s) + Q(s)$ to be Hurwitz, all quotients obtained must have positive coefficients.

of the above example using MATLAB will be

```
» a = [1 0 20 0 64];    % Num. coeff.vector
» b = [1 0 9 0];        % Den.coeff.vector
» [q,c] = deconv(a,b)   % Long division
q = 1 0                 % Quotient:  z(s)=q(1)*s+0
c = 0 0 11 0 64         % Coeff.of remainder polyn.
» c = c(3:5)            % Delete first two coeff.
c = 11 0 64
»[q,d] = deconv(b,c)    % Invert and divide
q = 0.0909 0            % Quotient:  y(s)=q(1)*s+0
d = 0 0 3.1818 0        % Coeff.of remainder polyn.
» d = d(3:4)            % Delete first two coeff.
d = 3.1818 0
» [q,e] = deconv(c,d)   % Invert and divide
q = 3.4571 0            % Quotient:  z(s)=q(1)*s+0
e = 0 0 64             % Coeff. of remainder polyn.
» e = e(3)              % Delete first two coeff.
e = 64
» [q,f] = deconv(d,e)   % Invert and divide
q = 0.0497 0            % Quotient:   y(s)=q(1)*s+0
f = 0 0                 % Remainder is zero
```

5.2.5   Removal of poles at the origin

If a lossless function has a pole at the origin, it can be removed by substraction. The remainder is

$$F_2(s) = F(s) - \frac{k_0}{s}$$

The order of $F_2(s)$ is one lower than that of $F(s)$. $F_2(s)$ is another lossless function with no pole at the origin.

Let's use the impedance function of (5.12).

$$Z_4(s) = Z(s) - \frac{\frac{64}{9}}{s} = \frac{s(s^2 + \frac{116}{9})}{s^2 + 9} \tag{5.15}$$

$$Y_5(s) = \frac{1}{Z_4(s)} - \frac{81}{116s} = \frac{\frac{35}{116}s}{s^2 + \frac{116}{9}}$$

And so on. Identifying each removal as the extraction of a network element also results in a network that realizes the given lossless function. The network of Fig. 5.5 also realizes the impedance function of (5.12).

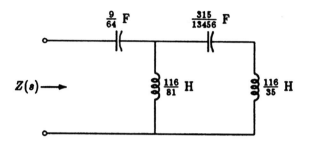

Figure 5.5: Network obtained by removing poles at the origin.

One convenient means of following what we are doing in this subsection is to replace every s with $1/p$. Then the origin in the s plane is mapped to the point at infinity in the p plane, and vice versa. Thus the removal of a pole at the origin in the s plane is equivalent to the removal of a pole at infinity in the p plane. Hence the algorithm in this subsection becomes completely analogous to that in the previous subsection if we simply visualize every $1/s$ as the variable.

The sequence of removal of poles at the origin can also be effected by long division as illustrated for this example in the following.

$$
\begin{array}{r}
\frac{64}{9}/s\ \Omega \\
9s + s^3 \overline{\smash{\big)}\ 64 + 20s^2 + s^4} \\
64 + \frac{64}{9}s^2 \\
\hline
\frac{116}{9}s^2 + s^4
\end{array}
$$

$$
\begin{array}{r}
\frac{81}{116}/s\ \mho \\
9s + s^3 \\
9s + \frac{81}{116}s^3 \\
\hline
\frac{35}{116}s^3
\end{array}
$$

$$
\begin{array}{r}
\frac{13456}{315}/s\ \Omega \\
\frac{116}{9}s^2 + s^4 \\
\frac{116}{9}s^2 \\
\hline
s^4
\end{array}
$$

$$
\begin{array}{r}
\frac{35}{116}/s\ \mho \\
\frac{35}{116}s^3 \\
\frac{35}{116}s^3 \\
\hline
\end{array}
$$

We could also use the continued-fraction notation to summarize the results of the long division.

$$F(s) = \frac{64}{9s} + \cfrac{1}{\cfrac{81}{116s} + \cfrac{1}{\cfrac{13,456}{315s} + \cfrac{1}{\cfrac{35}{116s}}}} \tag{5.16}$$

The realization by continued removal of poles at the origin is known as the *Cauer 2* realization.

The series of long divisions can be performed using the **deconv** command of MATLAB. For the above example, the MATLAB steps are

```
» a = [64 0 20 0 1];        % Num. coeff.vector
» b = [9 0 1 0];            % Den. coeff.vector
» [q,c] = deconv(a,b)
q = 7.1111 0                % Quotient:  z(s)=q(1)/s+0
c = 0 0 12.8889 0 1.0000
» c = c(3:5)
c = 12.8889 0 1.0000
» [q,d] = deconv(b,c)
q = 0.6983 0                % Quotient:  y(s)=q(1)/s+0
d = 0 0 0.3017 0
» d = d(3:4)
d = 0.3017 0
» [q,e] = deconv(c,d)
q = 42.7175 0              % Quotient:  z(s)=q(1)/s+0
e = 0 0 1
» e = e(3)
e = 1
» [q,f] = deconv(d,e)
q = 0.3017 0              % Quotient:  y(s)=q(1)/s+0
f = 0 0                   % Zero remainder
```

5.2.6   Removal of finite nonzero poles

If a lossless function has a pair of finite nonzero conjugate poles, they can be removed as well. For example, in (5.9)

$$F(s) = \frac{(s^2 + 1)(s^2 + 5)(s^2 + 20)}{s(s^2 + 2)(s^2 + 10)} \tag{5.17}$$

has a pair of poles at $s = \pm j\sqrt{2}$. The content of this pair of poles is given by the third term of its Foster's expansion in (5.10). To remove this pair of poles, we simply delete that term. Namely,

$$F_a(s) = F(s) - \frac{\frac{27}{8}s}{s^2 + 2} = \frac{5}{s} + s + \frac{\frac{45}{8}s}{s^2 + 10} = \frac{s^4 + \frac{165}{8}s^2 + 50}{s(s^2 + 10)} \quad (5.18)$$

This step can be accomplished by MATLAB as follows.

```
» Fs = '((s^2 + 1)*(s^2 + 5)*(s^2 + 20))/
(s*(s^2 + 2)*(s^2 + 10))';
» F1 = '(27*s)/(8*(s^2 + 2))';
» Fa = symsub(Fs,F1);
» simplify(Fa)
Fa = 1/8*(8*s^4 + 165*s^2 + 400)/s/(s^2 + 10)
```

Application of command `pretty(Fa)` yields

$$\frac{1}{8}\frac{8s^4 + 165s^2 + 400}{(s^2 + 10)s}$$

It should be obvious that such a removal leaves a remainder that is another lossless function, it retains all other poles, and its order is reduced by two. The zeros of the remainder are generally different from those of the original function.

5.2.7 Mixed canonic realization

The Foster and Cauer realizations are canonic in that the minimum number of elements are used to realize the network function. In the Foster realizations poles are separated and realized term by term. In Cauer realizations, one pole is removed (completely) at each step. Clearly, a network realizing a function cannot contain fewer elements than networks obtained by these methods.

Foster and Cauer realizations are not the only canonic realizations of a lossless function. We can realize a function by removing poles at various locations in any sequence we like as long as each removal is a complete one. For a moderately complex function, there can be a large number of possible canonic realizations. As an example, suppose we wish to realize the impedance

$$Z(s) = \frac{s(s^2 + 4)(s^2 + 25)}{(s^2 + 1)(s^2 + 9)} \ \Omega \tag{5.19}$$

We first remove its pole at infinity, leaving the remainder

$$Z_1(s) = \frac{s(s^2 + 4)(s^2 + 25)}{(s^2 + 1)(s^2 + 9)} - s = \frac{s(19s^2 + 91)}{(s^2 + 1)(s^2 + 9)} \ \Omega$$

Its reciprocal has a pole at the origin. We remove that pole next.

$$Y_2(s) = \frac{1}{Z_1(s)} - \frac{9}{91s} = \frac{s(91s^2 + 739)}{91(19s^2 + 91)} \ \mho$$

Its reciprocal has a pair of poles at $\pm j\sqrt{\frac{739}{91}}$. We remove them next, leaving

$$Z_3(s) = \frac{1}{Y_2(s)} - \frac{\frac{5760}{739}s}{s^2 + \frac{739}{91}} = \frac{8281}{739s} \ \Omega$$

Summarizing the successive removals, we obtain the network of Fig. 5.6.

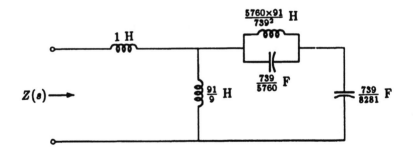

Figure 5.6: A canonic realization of the impedance function of (5.19).

Obviously, there are several possible canonic networks that will realize this same impedance, depending on the order in which poles are removed. As long as each removal of a pole or a pair of poles is complete, the network will still be canonic. Also, since each removal of pole(s) will alter the zeros of the remainder and these new zeros will become poles upon the reciprocation of a function, new poles that are quite unrelated to the original poles will be produced.

5.2.8 Noncanonic realization

Given a lossless function, if a part of a pole at the origin or infinity is removed, the remainder will be another lossless function of the same order. Likewise, if a part of a pair of conjugate poles is removed, the remainder will still be another lossless function of the same order. Thus if, at any stage in the pole removal sequence, the removal is not complete, a non-canonic realization results. Hence there are an infinite number of networks that will realize a given lossless function.

As an example, suppose we have an admittance

$$Y(s) = \frac{s(s^2 + 8)}{s^2 + 4} \tag{5.20}$$

which has a pole at infinity with $k_\infty = 1$. If we remove a part, say one-half, of this pole, the remainder is

$$Y_1(s) = \frac{s(s^2 + 8)}{s^2 + 4} - \frac{1}{2}s = \frac{s(s^2 + 12)}{2(s^2 + 4)}$$

The reciprocal of $Y_1(s)$ has a pole at the origin with $k_0 = 2/3$. Suppose we remove one half of this pole, the remainder is

$$Z_2(s) = \frac{1}{Y_1(s)} - \frac{1}{3s} = \frac{5s^2 + 12}{3s(s^2 + 12)}$$

The reciprocal of $Z_2(s)$ has a pair of poles at $\pm j\sqrt{\frac{12}{5}}$ and

$$\frac{1}{Z_2(s)} = \frac{3}{5}s + \frac{144s}{5(5s^2 + 12)}$$

We could remove a part of this pair of poles. Let

$$Y_3(s) = \frac{1}{Z_2(s)} - \frac{100s}{5(5s^2 + 12)} = \frac{s(3s^2 + 16)}{5s^2 + 12}$$

The function $Y_3(s)$ is still of the same order as $Y(s)$. We shall complete the realization by applying Foster 1 to $Y_3(s)$. We find

$$\frac{1}{Y_3(s)} = \frac{3}{4s} + \frac{11s}{4(3s^2 + 16)}$$

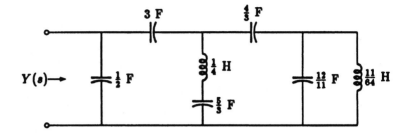

Figure 5.7: Network that realizes the admittance of (5.20).

The admittance of (5.20) is then realized by the network of Fig. 5.7.

At this juncture, we simply wish to demonstrate that the partial removal of poles can be done at will without reducing the complexity of the remainder of a lossless function. Hence by partial pole removal and reciprocation, we can move the zeros and poles along the j axis with great flexibility. Partial removal of poles happens to be a very important artifice in filter realization.

5.3   Properties of lossless twoports

A twoport with its port quantities shown in Fig. 5.8 can be characterized

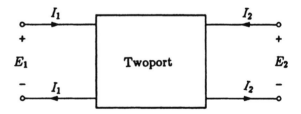

Figure 5.8: Notation used in a twoport.

by either its z matrix in the relationship

$$
\begin{bmatrix} E_1 \\ E_2 \end{bmatrix} = [z] \begin{bmatrix} I_1 \\ I_2 \end{bmatrix} = \begin{bmatrix} z_{11} & z_{12} \\ z_{21} & z_{22} \end{bmatrix} \begin{bmatrix} I_1 \\ I_2 \end{bmatrix}
\tag{5.21}
$$

or its y matrix in the relationship

$$\begin{bmatrix} I_1 \\ I_2 \end{bmatrix} = [y] \begin{bmatrix} E_1 \\ E_2 \end{bmatrix} = \begin{bmatrix} y_{11} & y_{12} \\ y_{21} & y_{22} \end{bmatrix} \begin{bmatrix} E_1 \\ E_2 \end{bmatrix} \tag{5.22}$$

Here, we shall limit our attention to reciprocal twoports - those containing inductors, capacitors, mutual inductances, and ideal transformers. The mathematical properties of the z matrix and those of the y matrix are exactly the same. We shall use the y matrix in our discussion.

It is obvious that the y_{11} and y_{22} of a lossless twoport must each be a lossless function. The transfer functions $y_{12} = y_{21}$ have properties that are slightly different from those of the lossless function. Specifically, each transfer admittance must have the following properties [Gu].

(1) It must be the ratio of an even polynomial to an odd polynomial, or vice versa.

(2) All its poles must lie on the j axis, they must be simple, their residues must be real (they may be positive or negative).

(3) Its zeros may lie anywhere in the s plane, but they must occur symmetrically about both axes. They may be of any order.

If we expand each of the elements of the y matrix of a lossless twoport into its Foster's form, we have the general expressions

$$y_{11}(s) = \frac{k_{110}}{s} + k_{11\infty}s + \sum_i \frac{k_{11i}s}{s^2 + \omega_i^2}$$

$$y_{21}(s) = \frac{k_{210}}{s} + k_{21\infty}s + \sum_i \frac{k_{21i}s}{s^2 + \omega_i^2} \tag{5.23}$$

$$y_{22}(s) = \frac{k_{220}}{s} + k_{22\infty}s + \sum_i \frac{k_{22i}s}{s^2 + \omega_i^2}$$

We must have

$$k_{11i}k_{22i} - k_{21i}^2 \geq 0 \tag{5.24}$$

where i may be zero, infinity, or finite nonzero. Equation (5.24) is known as the *residue condition*.

One consequence of the residue condition is that each pole that exists in $y_{21}(s)$ must also be present in both $y_{11}(s)$ and $y_{22}(s)$. The converse is not true. Poles that exist in $y_{11}(s)$ (or $y_{22}(s)$) but not in $y_{21}(s)$ are called the *private poles* of $y_{11}(s)$ (or $y_{22}(s)$).

5.4   LC ladder twoport

One of the most important classes of passive filters uses LC ladder networks. Many examples generated in Section 5.2 are already in ladder form and can be utilized as LC twoports. This is especially true when networks are generated by pole removal. For example, the network of Fig. 5.6 can be arranged as a twoport as shown in Fig. 5.9.

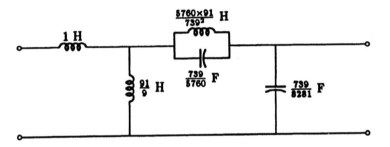

Figure 5.9: Network of Fig. 5.6 arranged as a twoport.

In the twoport of Fig. 5.9, it is clear that

$$z_{11}(s) = \frac{s(s^2 + 4)(s^2 + 25)}{(s^2 + 1)(s^2 + 9)} \tag{5.25}$$

The question is what is the z_{12} or z_{21} of the twoport. This can be answered to within a constant multiplier if we trace the history of the development of the ladder by pole removal.

Any time poles are removed completely from a driving-point function of a twoport, zeros at the locations of the poles are created in the transfer function of the twoport. This is because when impedance poles are removed, the removed circuits will look like open circuits in series at the pole frequencies. These circuits will not allow signals of those frequencies to be transmitted to the other port.

Similarly, when admittance poles are removed, the removed circuits look like short circuits in shunt at the pole frequencies. These circuits will not allow signals of those frequencies to be transmitted to the other port.

In Fig. 5.9, the first inductor *appears* to give z_{21} a zero at infinity as an inductance looks like an open circuit at high frequencies. However, this inductance does not contribute anything to z_{21}. It only gives rise to a private pole of z_{11}. Since $z_{21} = E_2/I_1|_{I_2=0}$, in obtaining z_{21} we are assuming that port 1 is excited by a current source. Hence the 1-henry inductor will contribute only to z_{11} but not to z_{21}.

The second inductor produces a zero at the origin, the parallel LC circuit (sometimes called a tank circuit) produces a pair of zeros at $\pm j\sqrt{\frac{739}{91}}$, and the last capacitor gives a zero at infinity to z_{21}. Hence all poles and zeros of z_{21} are known and we can expect that

$$z_{21}(s) = \frac{Ks\left(s^2 + \frac{739}{91}\right)}{(s^2 + 1)(s^2 + 9)} \tag{5.26}$$

The only remaining issue is the value of K in (5.26). To determine this value all we need to know is the value of z_{21} at one value of s. Frequently, $s = 0$, ∞, or 1 will be a good choice although, in principle, any value for s will be acceptable. For instance, let us use $s = 1$. The twoport has branch impedances shown in Fig. 5.10. Routine analysis will give

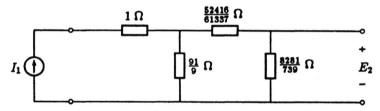

Figure 5.10: Branch impedances of the twoport of Fig. 5.9 at $s = 1$.

$$\left.\frac{E_2}{I_1}\right|_{s=1} = \frac{7553}{1478}$$

But (5.26) gives

$$z_{21}(1) = \frac{K\left(1 + \frac{739}{91}\right)}{20}$$

Equating the two quantities and solving for K, we get $K = \frac{8281}{739}$.

When poles are removed partially, it does not create zeros for the transfer function. This is because the remainder still has those poles. Hence at the pole frequencies, the network forms a voltage or current divider. A simple example will illustrate this observation.

Suppose $z_{11}(s) = \frac{s}{s^2+4}$. The admittance $1/z_{11}$ has a pole at infinity. We remove one-third of that pole, leaving

$$y_1 = \frac{s^2+4}{s} - \frac{1}{3}s = \frac{\frac{2}{3}s^2+4}{s}$$

The situation is shown in Fig. 5.11. Although the $\frac{1}{3}$-farad capacitor

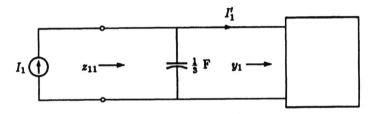

Figure 5.11: Partial removal of the admittance pole at infinity.

looks like a short circuit at high frequencies, $I_1' = \frac{2}{3}I_1$ as $s \to \infty$ because $y_1 = \frac{2}{3}s$ as $s \to \infty$.

As an example of the application of the partial removal of poles, let us start with

$$z_{11} = \frac{(s^2+1)(s^2+6)}{s(s^2+2)(s^2+10)} \ \Omega \qquad (5.27)$$

The admittance $1/z_{11}$ has a pole (equal to s) at infinity. We remove $\frac{14}{25}s$ from it, leaving

$$y_1 = \frac{1}{z_{11}} - \frac{14}{25}s = \frac{s(s^2+16)(11s^2+26)}{25(s^2+1)(s^2+6)}$$

which has a pair of zeros at $\pm j4$. The impedance $1/y_1$ will have a pair of poles there. We remove them completely, leaving

$$z_2 = \frac{1}{y_1} - \frac{\frac{25}{16}s}{s^2+16} = \frac{25(5s^2+6)}{16s(11s^2+26)}$$

The admittance $1/z_2$ has a pole at infinity (equal to $\frac{176}{125}s$). We remove $\frac{48}{37}s$ of that pole, leaving

$$y_3 = \frac{1}{z_2} - \frac{48}{37}s = \frac{512s(s^2+16)}{925(5s^2+6)}$$

The impedance $1/y_3$ will again have a pair of poles at $\pm j4$. We remove them completely, leaving

$$z_4 = \frac{1}{y_3} - \frac{34,225s}{4096(s^2+16)} = \frac{2775}{4096s}$$

Summarizing the successive removals and arranging the elements to form a twoport, the network of Fig. 5.12 results. This twoport will have the

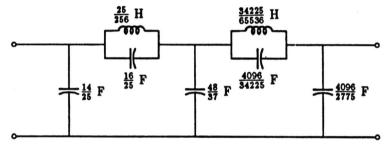

Figure 5.12: A twoport developed from the z_{11} of (5.27).

following transfer impedance:

$$z_{21}(s) = \frac{K(s^2+16)^2}{s(s^2+2)(s^2+10)}$$

To determine K, we investigate the twoport at $s = 0$. The twoport reduces to the situation shown in Fig. 5.13 in which $C = \frac{14}{25} + \frac{48}{37} + \frac{4096}{2775} = \frac{10}{3}$ F. Equating E_2/I_1 to $z_{21}(0) = \frac{16^2 K}{20s}$ we obtain $K = \frac{3}{128}$.

5.5   Foster's preamble

The driving-point function of a oneport that is not lossless will not be a lossless function. If the driving-point function has poles on the j axis, they must have real and positive residues. Hence these poles can be removed partially or completely, just as poles can be removed from a lossless function. If the purpose is simply to realize the driving-point

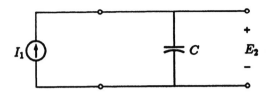

Figure 5.13: Behavior of twoport of Fig. 5.12 at $s = 0$.

function, the complete removal of poles simplifies the remainder. In some situations, particularly in filter realizations, this process can be applied repeatedly until the remainder is much simpler. This process is known as *Foster's preamble*.

For example, suppose we have an impedance

$$Z(s) = \frac{4s^4 + 21s^3 + 44s^2 + 30s + 40}{2(4s^3 + 21s^2 + 40s + 10)} \tag{5.28}$$

It has a pole at infinity. We remove it, leaving the remainder

$$Z_1 = Z(s) - \frac{1}{2}s = \frac{2(s^2 + 5s + 10)}{4s^3 + 21s^2 + 40s + 10}$$

Now, $1/Z_1$ has a pole at infinity. We remove it, leaving the remainder

$$Y_2 = \frac{1}{Z_1} - 2s = \frac{s^2 + 10}{2(s^2 + 5s + 10)}$$

The reciprocal of Y_2 has a pair of poles at $\pm j\sqrt{10}$. We can remove them (using (5.8) to evaluate the coefficient), leaving

$$Z_3 = \frac{1}{Y_2} - \frac{10s}{s^2 + 10} = 2$$

The impedance $Z(s)$ of (5.28) is realized by the network of Fig. 5.14.

As this example may suggest, if a driving-point function is that of a lossless ladder terminated in a resistance, it can be realized by applying Foster's preamble repeatedly. This is indeed the case if we know in advance the structure of the ladder. If we do not have this advance knowledge and there happens to have more than one Foster's preamble opportunity available at a juncture, there is no guarantee that the arbitrary removal of a pole or a pair of poles will leave a remainder on which

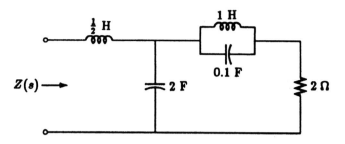

Figure 5.14: Realization of the impedance of (5.28).

the preamble can be repeated. In this situation, it may be necessary to explore various possibilities as to which one of these opportunities should be exercised first so that the process can be repeated as far as possible.

As an example, suppose we have the impedance function

$$Z_a(s) = \frac{(2s+1)(3s^2+1)}{s(6s^2+6s+5)} \tag{5.29}$$

which has a pole at the origin. If we remove that pole, the remainder will be

$$Z_b(s) = Z_a(s) - \frac{1}{5s} = \frac{30s^2+9s+4}{5(6s^2+6s+5)}$$

which has no j-axis poles or zeros and Foster's preamble is no longer applicable. On the other hand, $Z_a(s)$ also has a pair of zeros at $\pm j1/\sqrt{3}$. If we remove this pair of poles completely from $1/Z_a(s)$ first, the remainder will be

$$Y_c = \frac{1}{Z_a(s)} - \frac{3s}{3s^2+1} = \frac{2s}{2s+1}$$

which and has a zero at the origin. We can then remove the pole at the origin from $1/Y_c(s)$, leaving a remainder that is a 1- ohm resistor. Thus $Z_a(s)$ of (5.29) is realized by the circuit of Fig. 5.15.

Foster's preamble is a very useful technique in the realization of LC ladder filters. From the network functions of such filters, we can usually anticipate the structure of the ladder. Therefore, at each point of

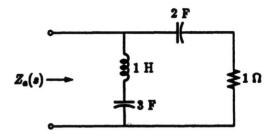

Figure 5.15: Realization of the impedance of (5.29).

the development of the ladder we know exactly what to expect and, therefore, apply Foster's preamble to the appropriate pole(s). Usually, there is little guesswork for the application of Foster's preamble for that purpose. This procedure will be taken advantage of in Chapter 7.

In the process of realizing the driving-point function of a terminated LC ladder, it is sometimes necessary to remove certain j-axis poles partially, much like the noncanonic realization of an unterminated LC ladder. Although such an operation is not, strictly speaking, one of the Foster's preamble steps, it is sometimes necessary to combine such an operation with Foster's preamble steps.

5.6 Summary

In this chapter, we have described certain passive network properties. Emphasis has been placed on the lossless networks, because most passive filters use this class of networks. We have also described certain synthesis techniques of passive networks. These topics form the basic foundation of the filter-realization procedures in the next two chapters.

What is included in this chapter is only a small fraction of the theory of passive networks and their synthesis. Our objective here has been to set out just enough background for the purposes of their application to filter realization. There are numerous interesting related topics, the thorough treatment of which is well beyond the scope of this text. The reader is encouraged to consult other references on the general and pure treatments of these advanced, more specialized topics.

Problems

5.1 For each of the following functions, determine whether or not it is lossless. If not, give your reasons.

(a) $\dfrac{s^2 + 25}{s(s^2 + 5)(s^2 + 50)}$

(b) $\dfrac{s(s^2 + 4)(s^2 + 9)}{(s^2 + 2)(s^2 + 10)}$

(c) $\dfrac{(s^2 + 5)(s^2 + 20)(s^2 + 100)}{s(s^2 + 10)(s^2 + 21)}$

(d) $\dfrac{s^3}{(s^2 + 2)(s^2 + 8)}$

5.2 Obtain the two Foster and the two Cauer realizations of the following admittance.

$$Y(s) = \frac{s^3 + 4s}{s^4 + 13s^2 + 30} \ \mho$$

5.3 Give at least two canonic realizations of the following impedance that are neither Foster nor Cauer realizations.

$$Z(s) = \frac{(s^2 + 2)(s^2 + 10)}{s(s^2 + 5)(s^2 + 12)} \ \Omega$$

5.4 An impedance, $Z(s)$, is obtained from the following circuit, which is a conanic network. Give the circuit structures of the four basic Foster and Cauer realizations of this impedance.

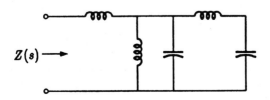

5.5 Realize the following impedance function by the network shown.

$$Z(s) = \frac{s(s^2 + 4)(s^2 + 25)}{(s^2 + 1)(s^2 + 9)(s^2 + 36)} \ \Omega$$

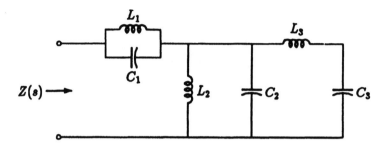

5.6 Realize the following impedance by first applying the Cauer 2 step once. Then complete the realization using Cauer 1 steps.

$$Z(s) = \frac{(s^2 + 1)(s^2 + 7)}{s(s^2 + 2)} \ \Omega$$

5.7 The pole-zero patterns of the three impedances, Z_1, Z_2, and Z_3, are shown below. The point at infinity has been brought 'in sight.' Give a reactance plot for each of the three combined impedances.

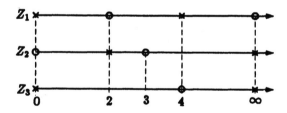

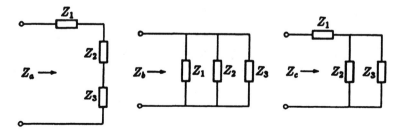

5.8 Determine for each of the following functions whether or not it can be the z_{12} of a lossless twoport. If not, state your reasons.

(a) $\dfrac{(s^2 + 9)(s^2 + 50)}{s(s^2 + 20)(s^2 + 25)}$

(b) $\dfrac{s^5}{(s^2 + 10)(s^2 + 30)}$

(c) $\dfrac{s^4 - 2s^2 + 12}{s(s^2 + 5)}$

(d) $\dfrac{s^4 + 2s^2 + 12}{s}$

(e) $\dfrac{s^2 + 50}{s(s^2 + 1)(s^2 + 2)}$

5.9 It is desired to remove a parallel capacitor from the impedance

$$\frac{(s^2 + 4)(s^2 + 10)}{s(s^2 + 9)(s^2 + 50)} \; \Omega$$

such that the remainder will have zeros at $\pm j8$. Determine the capacitance of the capacitor.

5.10 Find an LC ladder to realize the following impedances simultaneously. K is arbitrary. Find the value of K for your network.

$$z_{11} = \frac{s^2 + 5}{s(s^2 + 10)} \qquad z_{21} = \frac{Ks^2}{s(s^2 + 10)}$$

5.11 Find an LC ladder to realize the following impedances simultaneously. K is arbitrary. Find the value of K for your network.

$$y_{11} = \frac{s^2 + 2}{s(s^2 + 5)} \qquad y_{21} = \frac{Ks}{(s^2 + 5)}$$

5.12 Realize the following impedance function by applying Foster's preamble.

$$Z(s) = \frac{6s^3 + 30s^2 + 5s}{3s^2 + 12s + 2} \; \Omega$$

5.13 Realize the following admittance function by applying Foster's preamble.

$$Y(s) = \frac{s(28s^2 + 12s + 1)}{40s^4 + 24s^3 + 33s^2 + 15s + 1} \; \mho$$

Chapter 6

Singly-terminated LC ladders

We are now in a position to realize a very useful configuration, one of the simplest arrangements in filter applications. We shall deal with the situation where the source is an ideal one - either a voltage or a current source. The filter itself is an LC ladder. The load is a pure resistance.

6.1 LC ladder with a current source

The arrangement of Fig. 6.1 is our starting point. The load has been normalized to be 1 ohm for convenience. Using the twoport impedance parameters defined in (5.21) for the LC twoport, we can write

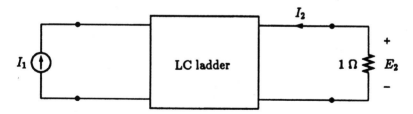

Figure 6.1: Singly-terminated filter with a current source.

$$E_2 = z_{21}I_1 + z_{22}I_2 = z_{21}I_1 - z_{22}E_2$$

Hence

$$Z_{21}(s) = \frac{E_2}{I_1} = \frac{z_{21}}{1 + z_{22}} \tag{6.1}$$

It is helpful to discuss first the zeros of the transfer function $Z_{21}(s)$. Based on our discussion in Chapter 5 in regard to the properties of the z parameters of an LC twoport and (6.1), we can state that *the zeros of $Z_{21}(s)$ are the zeros of $z_{21}(s)$ and the private poles of $z_{22}(s)$.* This is because if z_{21} is zero, $Z_{21}(s)$ will also be zero. Also, if z_{22} is infinite and z_{21} is finite, $Z_{21}(s)$ will be zero. We shall call the zeros of any transfer function the *transmission zeros*.

Since the $z_{21}(s)$ of an LC twoport must be the ratio of an even polynomial to an odd polynomial, or vice versa, and since z_{22} must be a lossless function, $Z_{21}(s)$ must be of the form

$$Z_{21}(s) = \frac{P(s)}{Q(s)} = \frac{KN_1(s)}{M_2(s) + N_2(s)} \quad \text{or} \quad \frac{KM_1(s)}{M_2(s) + N_2(s)} \tag{6.2}$$

in which each M represents the even part of a polynomial and each N the odd part. Further, $Q(s)$ must be a Hurwitz polynomial.

To be specific, if $P(s)$ is an odd polynomial ($M_1 = 0$), we can divide both the numerator and denominator by $M_2(s)$ to form

$$Z_{21}(s) = \frac{\dfrac{KN_1(s)}{M_2(s)}}{1 + \dfrac{N_2(s)}{M_2(s)}} \tag{6.3}$$

We can then identify

$$z_{22} = \frac{N_2(s)}{M_2(s)} \quad \text{and} \quad z_{21} = \frac{KN_1(s)}{M_2(s)} \tag{6.4}$$

On the other hand, if $P(s)$ is an even polynomial ($N_1 = 0$), we can divide both the numerator and denominator by $N_2(s)$ to form

$$Z_{21}(s) = \frac{\dfrac{KM_1(s)}{N_2(s)}}{1 + \dfrac{M_2(s)}{N_2(s)}} \tag{6.5}$$

We can then identify

$$z_{22} = \frac{M_2(s)}{N_2(s)} \quad \text{and} \quad z_{21} = \frac{KM_1(s)}{N_2(s)} \tag{6.6}$$

For a given $Z_{21}(s)$, now it's a matter of realizing an LC network such that (6.4) or (6.6) is satisfied. We have all the basic tools for carrying out this realization process as long as the zeros of $Z_{21}(s)$ lie on the j axis, including the origin and infinity.

One remark is appropriate here. Typical examples in Chapter 5 realize z_{11} and z_{21} (to within a constant multiplier) simultaneously. To satisfy (6.4) or (6.6), we need to realize z_{22} and z_{21} simultaneously. Hence the orientation is to realize the driving-point at port 2 and, at the same time, realize the zeros of the transfer function. This, of course, does not involve any new principle. We merely have to change our direction of approach - instead of proceeding from the left to the right, we now go from the right to the left.

6.1.1 Transmission zeros at the origin and infinity

When a filter has only transmission zeros at the origin and at infinity, the procedure for realizing it is particularly simple. This is because Cauer's realization steps removes poles at the origin or infinity completely. A few examples should illustrate what these realizations entail.

EXAMPLE 6.1. Realize the third-order Butterworth lowpass filter for the singly-terminated arrangement of Fig. 6.1.

SOLUTION We have

$$Z_{21}(s) = \frac{K}{s^3 + 2s^2 + 2s + 1} \tag{6.7}$$

Since the numerator is even, we write (6.7) as

$$Z_{21}(s) = \frac{\dfrac{K}{s^3 + 2s}}{1 + \dfrac{2s^2 + 1}{s^3 + 2s}}$$

There are three transmission zeros, all at $s = \infty$. Since each Cauer 1 step removes a pole at infinity completely, we apply Cauer 1 steps three times to

$$z_{22} = \frac{2s^2 + 1}{s^3 + 2s}$$

We now perform the long division

$$
\begin{array}{r}
\frac{1}{2}s \; \mho \\
2s^2 + 1 \; \overline{\big)\; s^3 + 2s}
\end{array}
$$

$$
\begin{array}{r}
s^3 + \frac{1}{2}s \quad \frac{4}{3}s \; \Omega \\
\hline
\frac{3}{2}s \; \big)\; 2s^2 + 1 \\
2s^2 \qquad \frac{3}{2}s \; \mho \\
\hline
1 \; \big)\; \frac{3}{2}s \\
\frac{3}{2}s \\
\hline
\end{array}
$$

and the network of Fig. 6.2 is obtained.

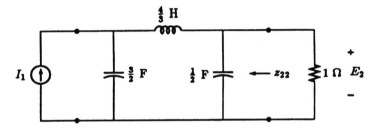

Figure 6.2: Network that realizes the Z_{21} of (6.7).

To evaluate K, we see that at $s = 0$,

$$
\frac{E_2}{I_1} = 1 = Z_{21}(0) = K
$$

Hence $K = 1$.

EXAMPLE 6.2. Realize the highpass Butterworth filter with

$$
Z_{21}(s) = \frac{Ks^3}{s^3 + 2s^2 + 2s + 1} \tag{6.8}
$$

SOLUTION   Write

$$
Z_{21}(s) = \frac{\dfrac{Ks^3}{2s^2 + 1}}{1 + \dfrac{s^3 + 2s}{2s^2 + 1}}
$$

We have three transmission zeros at the origin. Since each Cauer 2 step removes a pole at the origin completely, we can simply apply Cauer 2 steps three times to

$$z_{22} = \frac{s^3 + 2s}{2s^2 + 1}$$

Performing long division, we get

$$
\begin{array}{r}
\frac{1}{2}/s\ \mho \\
2s + s^3 \enclose{longdiv}{1 + 2s^2} \\
\end{array}
$$

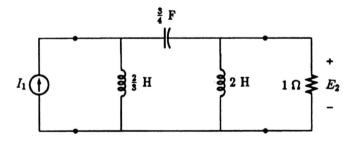

Summarizing the results of the long division, we obtain the network of Fig. 6.3.

To determine K, we see that at $s = \infty$,

$$\frac{E_2}{I_1} = 1 = Z_{21}(\infty) = K$$

Hence $K = 1$.

Figure 6.3: Network that realizes the $Z_{21}(s)$ of (6.8).

EXAMPLE 6.3. Realize the transfer impedance

$$Z_{21}(s) = \frac{Ks}{s^3 + 2s^2 + 2s + 1} \tag{6.9}$$

SOLUTION   We write

$$Z_{21}(s) = \cfrac{\cfrac{Ks}{2s^2 + 1}}{1 + \cfrac{s^3 + 2s}{2s^2 + 1}} \tag{6.10}$$

From (6.9) we see that there are two transmission zeros at infinity and one at the origin. From (6.10) we see that

$$z_{22} = \frac{s^3 + 2s}{2s^2 + 1}$$

has a private pole at infinity. Let us remove this private pole first, leaving

$$z_1 = z_{22} - \frac{1}{2}s = \frac{3s}{4s^2 + 2}$$

The reciprocal of z_1 is

$$y_1 = \frac{1}{z_1} = \frac{4s^2 + 2}{3s} = \frac{4}{3}s + \frac{2}{3s}$$

which is just right for realizing one transmission zero at the origin and another one at infinity. The network is shown in Fig. 6.4.

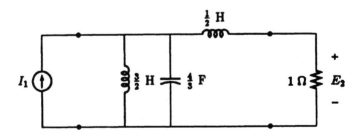

Figure 6.4: Network that realizes the $Z_{21}(s)$ of (6.9).

To evaluate K, we let $s \to 0$. The circuit approaches that of Fig. 6.5, in which

$$\frac{E_2}{I_1} = \frac{I_L}{I_1} = \left.\frac{\frac{3}{2}s}{\frac{3}{2}s + 1}\right|_{s \to 0} = \frac{3}{2}s = Z_{21}(0) = Ks$$

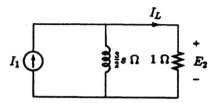

Figure 6.5: Asymptotic behavior of the network of Fig. 6.4 as $s \to 0$.

Hence $K = \frac{3}{2}$.

Alternatively, we may take the same z_{22} and remove poles in a different sequence. Let us remove a pole at the origin first, followed by the removal of two poles at infinity. This can be accomplished by the following long division.

$$
\begin{array}{r}
\frac{1}{2}/s\ \mho \\
2s + s^3\ \overline{\big|\ 1 + 2s^2} \\
\underline{1 + \frac{1}{2}s^2} \qquad \frac{2}{3}s\ \Omega \\
\frac{3}{2}s^2\ \big|\ s^3 + 2s \\
\underline{s^3} \qquad \frac{3}{4}s\ \mho \\
2s\ \big|\ \frac{3}{2}s^2 \\
\underline{\frac{3}{2}s^2}
\end{array}
$$

The result is another circuit that is shown in Fig. 6.6.

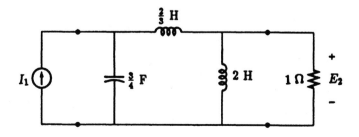

Figure 6.6: Another network that realizes the $Z_{21}(s)$ of (6.9).

To evaluate K, we let $s = 1$ for the sake of variety. The impedances of the elements are shown in Fig. 6.7. We first assume that $E_2 = 1$ V. The various quantities can then be obtained successively from the right to the left, resulting in

$$I_1 = 3 \text{ A}$$

Hence

$$\left. \frac{E_2}{I_1} \right|_{s=1} = \frac{1}{3} = Z_{21}(1) = \frac{K}{6}$$

Hence $K = 2$.

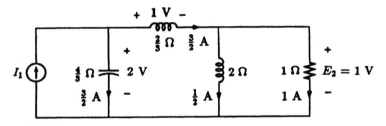

Figure 6.7: Impedances of the branches of the network of Fig. 6.6 at $s = 1$.

A few observations can be made based on the examples above.

(1) As long as transmission zeros are located at the origin or infinity, all that is needed is the application of Cauer 1 or 2 steps to remove poles from either the admittance or the impedance at the origin or infinity. These removals can always be effected by long division.

(2) Although we can produce the necessary transmission zeros by this procedure, we have no control over the constant multiplier. The multiplier can always be evaluated by examining the performance of network at one value of s.

(3) The order in which the poles are removed is arbitrary. Except for very simple networks or when all transmission zeros occur at the origin or infinity, generally there are several possible circuits that

will realize a given $Z_{21}(s)$, depending on the order in which the poles are removed. The proportionality constants are generally different for different realizations.

(4) In realizing a $Z_{21}(s)$, it is implicit that the source is a current source. Hence, the last element in the development of z_{22} should always be a shunt one.

6.1.2 Zero shifting

In Chapter 5, we have demonstrated that the partial removal of poles from a function does not affect the order and the poles of the remainder, but it does affect the zeros of the latter. We will now take a closer look at this procedure in order to place transmission zeros at desired locations as our filters may demand. We shall illustrate the procedure with an example. Suppose we have an impedance

$$Z(s) = \frac{(s^2 + 1)(s^2 + 9)}{s(s^2 + 4)} \tag{6.11}$$

which has a pole at the origin, one at infinity, and a pair at $\pm j2$. The effect of pole removal on the zeros is best seen by observing the plots of the various reactance functions.   In Fig. 6.8, we have a plot of the reactance function $\text{Im}[Z(j\omega)] = X(\omega)$ for the impedance of (6.11). If we remove part of the pole at infinity (up to $k_{\infty}s$ Ω), a reactance equal to $k'_{\infty}\omega$ ($k'_{\infty} < k_{\infty}$) is subtracted from $X(\omega)$ as shown. The zeros of the

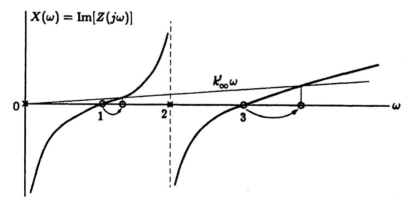

Figure 6.8: Effects on zeros when part of the pole at infinity is removed.

remainder will be located where $X(\omega)$ and $k'_{\infty}\omega$ intersect. It is seen that both zeros shift toward $\omega = \infty$.

If part of the pole at the origin is removed (up to k_0/s), we are subtracting $-k_0'/\omega$ (where $k_0' < k_0$) from $X(\omega)$. This is depicted in Fig. 6.9. The

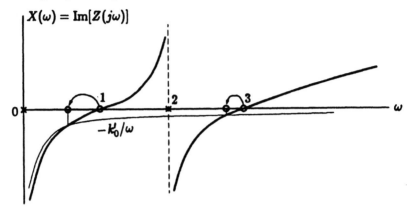

Figure 6.9: Effect on zeros when part of the pole at the origin is removed.

zeros of the remainder will be located at the intersection of $X(\omega)$ and $-k_0'/\omega$. The zeros are moved toward the origin.

If part of the poles at $\pm j2$ is removed (up to $k_1 s/(s^2 + 4)$), a reactance equal to $k_1'\omega/(4-\omega^2)$ ($k_1' < k_1$) is subtracted from $X(\omega)$. This is depicted in Fig. 6.10. It is seen that the zeros of the remainder shift toward $\pm j2$.

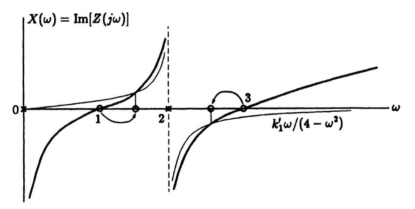

Figure 6.10: Effects on zeros when part of finite poles is removed.

An additional flexibility is afforded by taking the reciprocal of $Z(s)$ and then removing part of its poles, as illustrated in Fig. 6.11. Such a step is particularly useful if zeros near (but not equal to) $\pm j2$ are desired.

From the above examples, several conclusions can be drawn.

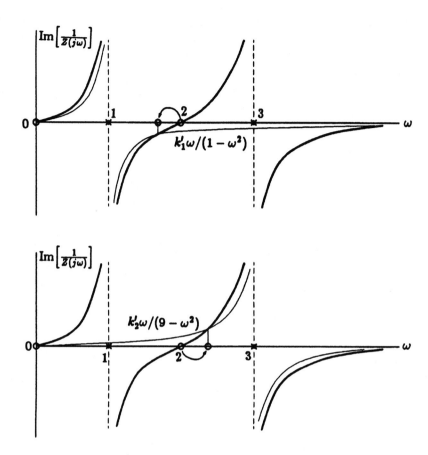

Figure 6.11: Partial removal of poles from the reciprocal of a function.

(1) A partial pole removal shifts all zeros (except those located at the origin or infinity) toward the affected pole(s).

(2) The larger the fraction of a pole or pole pair is removed, the more the zeros are shifted.

(3) Zeros cannot be shifted beyond the poles adjacent to them.

6.1.3 LC ladder with finite nonzero transmission zeros

Zero shifting can be used in filter realization to produce finite nonzero transmission zeros. When finite nonzero transmission zeros are desired in a filter (as would be the case in an elliptic filter) and the particular $z_{22}(s)$ does not have poles or zeros at these points, we can apply partial pole removal to shift zeros to these points. The reciprocal of the remainder will then have poles there. When these new poles are removed completely, transmission zeros are created there.

EXAMPLE 6.4. Realize

$$Z_{21}(s) = \frac{K(s^2 + 4)}{s^3 + 2s^2 + 2s + 1} \tag{6.12}$$

SOLUTION   We wish to realize

$$z_{22} = \frac{2s^2 + 1}{s^3 + 2s} \tag{6.13}$$

such that transmission zeros are produced at $\pm j2$ and infinity. Suppose we wish to produce transmission zeros at $\pm j2$ first, that is we want to shift a pair of zeros to $\pm j2$. A sketch of $\text{Im}[z_{22}(j\omega)]$ is shown in Fig. 6.12. It is easy to see that it is not possible to shift any zeros to $\pm j2$ by the partial removal of any pole of z_{22}.

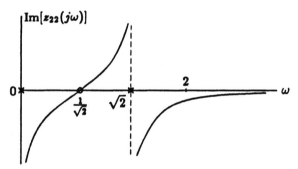

Figure 6.12: A plot of the reactance function of (6.13).

We next try the reciprocal of z_{22}. The susceptance function of $1/z_{22}(j\omega)$ is shown in Fig. 6.13. We can now remove part of its pole at infinity to shift the zeros at $\pm j\sqrt{2}$ to $\pm j2$. The value of C_1 can be determined by making

$$\left[\frac{1}{z_{22}} - C_1 s\right]_{s=j2} = \left[\frac{s^3 + 2s}{2s^2 + 1} - C_1 s\right]_{s=j2} = 0$$

Solving, we obtain $C_1 = 2/7$ F. The remainder is

$$y_1 = \frac{1}{z_{22}} - \frac{2}{7}s = \frac{3s(s^2 + 4)}{7(2s^2 + 1)}$$

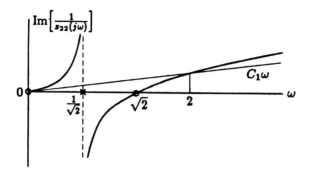

Figure 6.13: The susceptance function and partial removal of its pole at infinity.

Now the poles at $\pm j2$ can be removed from $1/y_1$ leaving

$$z_2 = \frac{1}{y_1} - \frac{\frac{49}{12}s}{s^2 + 4} = \frac{7}{12s}$$

The final remainder gives an admittance pole at infinity, which is the desired third transmission zero. The successive removals are summarized in Fig. 6.14.

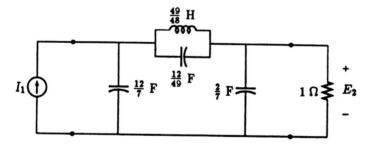

Figure 6.14: LC ladder with transmission zeros at $\pm j2$ and infinity.

To determine K, we let $s \to 0$, for which

$$\frac{E_2}{I_1} = 1 = Z_{21}(0) = 4K$$

Hence, $K = 1/4$.

6.2 LC ladder with a voltage source

If the source in a singly-terminated twoport is a voltage source as shown
in Fig. 6.15, we shall treat I_2 as the output. This is done to give a
complete theoretical duality to the case when the source is a current
source as treated in the previous section. Since the load is normalized
to 1 ohm, $E_2 = -I_2$ and $I_2/E_1 = -E_2/E_1$. Using the second equation
implied by (5.22), the pertinent relationship is

$$I_2 = y_{21}E_1 + y_{22}E_2 = y_{21}E_1 - y_{22}I_2 \tag{6.14}$$

which gives

$$Y_{21}(s) = \frac{I_2}{E_1} = \frac{y_{21}}{1 + y_{22}} \tag{6.15}$$

Since the y parameters and the z parameters of a lossless twoport have
the same mathematical properties, the procedure for realizing a given
$Y_{21}(s)$ is exactly dual to that for realizing a given $Z_{21}(s)$.

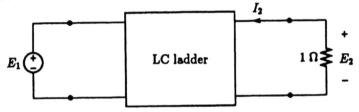

Figure 6.15: Singly-terminated filter with a voltage source.

EXAMPLE 6.5. Realize the transfer admittance

$$Y_{21}(s) = \frac{Ks(s^2 + 9)}{s^4 + s^3 + 5s^2 + 2s + 4} \tag{6.16}$$

SOLUTION   We write

$$Y_{21}(s) = \frac{\dfrac{Ks(s^2 + 9)}{s^4 + 5s^2 + 4}}{1 + \dfrac{s^3 + 2s}{s^4 + 5s^2 + 4}}$$

$$y_{22} = \frac{s^3 + 2s}{s^4 + 5s^2 + 4} \qquad\qquad y_{21} = \frac{Ks(s^2 + 9)}{s^4 + 5s^2 + 4}$$

From (6.16), we see that there are four transmission zeros - one at the origin, one at infinity, and a pair at $\pm j3$. The ones at the origin and infinity can easily be realized by removing the poles there from $1/y_{22}$, leaving

$$z_1 = \frac{1}{y_{22}} - s - \frac{2}{s} = \frac{s}{s^2 + 2}$$

Next we need to shift the zeros of $1/z_1$ to $\pm j3$. This can be accomplished by removing a part of the pole at infinity from $1/z_1$. We set

$$y_2(j3) = \left[\frac{1}{z_1} - C_1 s\right]_{s=j3} = 0$$

Solution yields $C_1 = 7/9$. Thus

$$y_2 = \frac{1}{z_1} - \frac{7}{9}s = \frac{2(s^2 + 9)}{9s} = \frac{2}{9}s + \frac{2}{s}$$

Summarizing the removals, we get the network of Fig. 6.16.

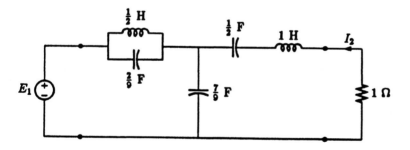

Figure 6.16: Network realizing the $Y_{21}(s)$ of (6.16).

To evaluate K, let us examine the network as $s \to 0$. The circuit approaches that shown in Fig. 6.17, in which

$$\frac{I_2}{E_1} = \frac{-1}{1 + \frac{2}{s}}\bigg|_{s \to 0} = \frac{-s}{2} = Y_{21}(0) = \frac{K9s}{4}$$

Hence $K = -\frac{2}{9}$.

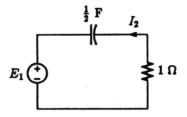

Figure 6.17: Asymptotic behavior of the circuit of Fig. 6.16 as $s \to 0$.

6.3   Other singly-terminated orientations

We have treated the arrangements in which sources are ideal - no internal impedance. This is the idealized version of some important practical situations. An ideal current source could represent the output of a circuit that can best be characterized as a Norton branch with a very high equivalent impedance. A practical example is the output of a field-effect transistor amplifier.

The ideal voltage source could represent a circuit whose output could be represented by its Thévenin branch with a negligible equivalent impedance. An example is the output of an amplifier with an op amp connected at the output terminals.

The arrangement of Fig. 6.1 can also be applied to the current ratio since $I_2/I_1 = -E_2/I_1 = -Z_{21}(s)$. In addition, if we apply the reciprocity theorem to the two ports, the arrangement of Fig. 6.18(a) results. We can then regard I_2' as the excitation and E_1' as the response and $E_1'/I_2' = E_2/I_1 = Z_{21}(s)$. Converting the Norton branch connected to port 2 into its Thévenin equivalent, we obtain the arrangement of Fig. 6.18(b). We then have $E_1'/E_2' = E_1'/I_2' = Z_{21}(s)$.

The arrangements of Fig. 6.18 can also be important when the source impedance is finite and the output is connected to a device or network whose impedance is so high that it is essentially an open circuit. Examples of these are the inputs of many amplifiers or op amp circuits.

The arrangement of Fig. 6.15 also applies to the voltage ratio $E_2/E_1 = -I_2/E_1 = -Y_{21}(s)$, as was pointed out in Section 6.2. Also, we can modify this arrangement to accommodate some other situations. If we apply the reciprocity theorem to the two ports, the arrangement of Fig. 6.19(a) results. We can regard E_2'' as the excitation and I_1'' as the response. Reciprocity assures that $I_1''/E_2'' = I_2/E_1 = Y_{21}(s)$. If we further replace the Thévenin branch at port 2 with its Norton equivalent, the arrangement of Fig. 6.19(b) results, in which $I_1''/I_2'' = I_1''/E_2'' = Y_{21}(s)$. The arrange-

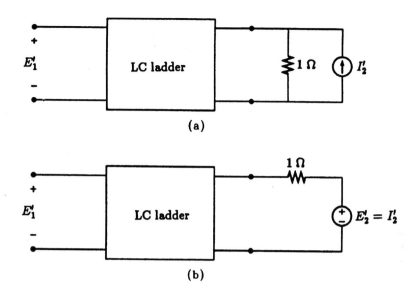

Figure 6.18: Arrangements derived from Fig. 6.1.

ments of Fig. 6.19 are meaningful when the source impedance is finite and the load impedance is very low in comparison.

6.4 Summary

In this chapter, we have treated in detail the filter applications in which only one finite resistive termination is present. The resistance may be the internal impedance of a source or it may be that of a load (which may represent the input impedance of the next circuit.) In every case, only one resistance is present.

We have also limited our filters to LC ladders. This means that the filters can have only j-axis transmission zeros. When the transmission zeros are all located at the origin and/or infinity, only Cauer steps of pole removal are required. When finite j-axis zeros are desired, zero shifting may become necessary. Fortunately, most filtering requirements can be satisfied by using LC ladders.

When some of the transmission zeros are not on the j axis, special realization techniques are required. These techniques are beyond the scope of this text.

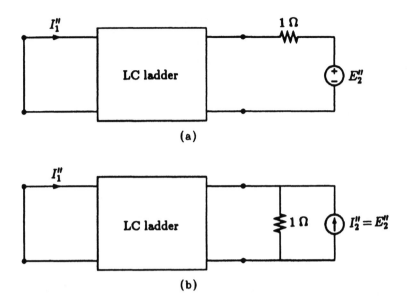

Figure 6.19: Arrangement derived from Fig. 6.15.

Problems

6.1 For the arrangement of Fig. 6.1, realize an LC ladder to give the following transfer impedances $Z_{21}(s)$. In each case, evaluate K.

(a) $\dfrac{Ks}{s^2 + 4s + 2}$

(b) $\dfrac{K}{s^3 + 5s^2 + 4s + 10}$

(c) $\dfrac{Ks}{s^3 + 2s^2 + 3s + 1}$

(d) $\dfrac{Ks^2}{s^3 + 2s^2 + 3s + 2}$

(e) $\dfrac{Ks^3}{s^3 + 5s^2 + 7s + 10}$

6.2 For the arrangement of Fig. 6.1, realize all possible LC ladders using the minimum number of elements to give each of the following transfer impedances $Z_{21}(s)$. Evaluate K for each network.

(a) $\dfrac{Ks}{s^3 + s^2 + 9s + 4}$

(b) $\dfrac{Ks^2}{s^3 + 3s^2 + 4s + 2}$

(c) $\dfrac{Ks^2}{s^4 + s^3 + 4s^2 + 2s + 3}$

6.3 For the arrangement of Fig. 6.15, realize an LC ladder to give the following transfer admittances $Y_{21}(s)$. In each case, evaluate K.

(a) $\dfrac{K}{s^3 + 5s^2 + 6s + 3}$

(b) $\dfrac{Ks}{s^3 + s^2 + 8s + 3}$

(c) $\dfrac{Ks^2}{s^3 + 2s^2 + 4s + 3}$

(d) $\dfrac{Ks^3}{s^3 + 4s^2 + 5s + 2}$

6.4 For the arrangement of Fig. 6.15, realize all possible LC ladders to give the following transfer admittance $Y_{21}(s)$.

$$\frac{Ks^2}{30s^4 + 40s^3 + 23s^2 + 5s + 1}$$

6.5 Remove an appropriate branch from $Z(s)$ or its reciprocal such that the remainder is another lossless function with zeros at $\pm j$. Also, obtain the remainder.

$$Z(s) = \frac{s(s^2 + 16)}{(s^2 + 4)(s^2 + 25)}$$

6.6 Remove an appropriate branch from $Z(s)$ or its reciprocal such that the remainder is another lossless function with zeros at $\pm j2$.

$$Z(s) = \frac{s(s^2 + 9)}{(s^2 + 1)(s^2 + 25)}$$

6.7 Remove an appropriate branch from $Y(s)$ or its reciprocal such that the remainder is another lossless function with zeros at $\pm j3$. Obtain the remainder.

$$Y(s) = \frac{s(s^2 + 12)}{(s^2 + 5)(s^2 + 20)}$$

6.8 It is desired to remove a single branch from $Z(s)$ or its reciprocal such that the remainder has zeros at $\pm j3$ and is still a lossless function. There are two different ways of accomplishing this. Perform both removals and, in each case, obtain the remainder.

$$Z(s) = \frac{s(s^2 + 4)(s^2 + 36)}{(s^2 + 1)(s^2 + 12)}$$

6.9 Obtain an LC ladder to realize the following impedance functions simultaneously. Evaluate K.

$$z_{22} = \frac{s^2 + 5}{s(s^2 + 10)} \qquad z_{21} = \frac{Ks}{(s^2 + 10)}$$

6.10 Obtain an LC ladder to realize the following admittance functions simultaneously. Evaluate K.

$$y_{22} = \frac{4s^2 + 3}{s^3 + 6s} \qquad y_{21} = \frac{K(s^2 + 16)}{s^3 + 6s}$$

6.11 Obtain an LC ladder which, when terminated in a 1-ohm resistance, will give the transfer impedance

$$Z_{21}(s) = \frac{K(s^2 + 9)}{s^3 + 2s^2 + 2s + 1}$$

Evaluate K.

6.12 Obtain an LC ladder which, when terminated in a 1-ohm resistance, will give the voltage ratio

$$\frac{E_2}{E_1} = \frac{K(s^2 + 8)}{s^3 + 2s^2 + 4s + 3}$$

Evaluate K.

6.13 Obtain an LC ladder which, when terminated in a 1-ohm resistance, will give the voltage ratio

$$\frac{E_2}{E_1} = \frac{K(s^2 + 2)}{s^3 + 3s^2 + 4s + 2}$$

Evaluate K.

6.14 Realize the following transfer function as the transfer admittance of a twoport terminated in a 1-ohm resistance. Use as few elements as possible.

$$Y_{21} = \frac{K(s^4 + 25s^2 + 100)}{s^5 + s^4 + 13s^3 + 6s^2 + 30s + 5}$$

Evaluate K.

6.15 The following transfer function gives a lowpass characteristic with equal-ripple variation inside both the pass band and the stop band. $\alpha_p = 1$ dB, $\alpha_s = 55$ dB, $\omega_p = 1$, and $\omega_s = 2.16199$. Realize it as the current ratio of a twoport terminated in a 1-ohm resistance. Evaluate K.

$$H(s) = \frac{K(s^2 + 5.38434)(s^2 + 28.859109)}{s^4 + 0.946621s^3 + 1.47823s^2 + 0.770255s + 0.310038}$$

Chapter 7

Doubly-terminated LC ladders

The filter configurations dealt with in Chapter 6 assume that one of the ports of the lossless twoport is restricted to one of the idealized conditions - open circuit, short circuit, ideal voltage source, or ideal current source. Another situation that is often encountered in filter applications is when finite impedances exists at both ports. This filter arrangement is often referred to as the *doubly-terminated lossless network*. The lossless twoport is often referred to as the *insertion network* or *insertion filter*. The theoretical basis and design technique for this configuration is quite different from the singly-terminated configuration.

7.1 Basic formulation

The doubly-terminated filter arrangement can be represented by the circuit of Fig. 7.1. The E_s and R_1 combination represents the Thévenin equivalent of the circuit to the left of port 1. The resistance R_2 represents the impedance of a load, which may represent the input impedance of the next circuit.

In the doubly-terminated situation, the voltage or current ratio (or its equivalent) is no longer meaningful. Rather, the performance of the filter should be measured by the amount of power delivered to the load compared with the maximum power deliverable by the source. It is well known that maximum power is delivered to the load when the load impedance and the source impedance are conjugate of each other. When these impedances are represented by resistances, the maximum power is delivered when the two resistances are equal. Therefore, if $R_1 = R_2$,

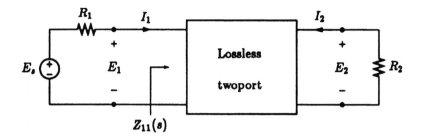

Figure 7.1: Arrangement of a doubly-terminated lossless twoport.

maximum power is delivered to the load when R_2 is connected to the source equivalent circuit. If $R_1 \neq R_2$, maximum power is delivered if an ideal transformer is used as shown in Fig. 7.2 so that the two resistances appear to be matched with each other. This is the optimal situation with which filter performance is compared in the doubly-terminated situation.

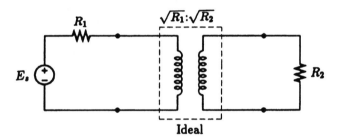

Figure 7.2: Arrangement in which maximum power is delivered to the load.

The maximum power deliverable from the source is

$$P_{\text{max}} = \frac{|E_s|^2}{4R_1} \tag{7.1}$$

The power delivered to the load is

$$P_2 = \frac{|E_2|^2}{R_2} \tag{7.2}$$

We shall define a function called the *transmission coefficient*, denoted by $t(s)$, which, for $s = j\omega$, gives the power ratio

$$|t(j\omega)|^2 = \frac{P_2}{P_{\max}} \tag{7.3}$$

From (7.1) and (7.2), we have

$$|t(j\omega)|^2 = \frac{4R_1}{R_2}\left|\frac{E_2}{E_s}\right|^2 \tag{7.4}$$

It is obvious that

$$|t(j\omega)|^2 \le 1 \tag{7.5}$$

In filtering, we wish to have $|t(j\omega)|$ approach unity in the pass band and approach zero in the stop band. When $|t(j\omega)|$ is nearly unity, then almost all available power is transmitted to the load. On the other hand, when $|t(j\omega)|$ is nearly zero, very little of the available power is transmitted to the load. We can view this latter condition to be the result of the fact that most of the available power is reflected back to the source. Therefore, it is useful to define another function called the *reflection coefficient*, denoted by $\rho(s)$, such that for $s = j\omega$,

$$|\rho(j\omega)|^2 = 1 - |t(j\omega)|^2 \tag{7.6}$$

Next, we shall denote the input impedance at port 1 with port 2 terminated in R_2 as $Z_{11}(s)$. For $s = j\omega$,

$$Z_{11}(j\omega) = R_{11}(\omega) + jX_{11}(\omega) \tag{7.7}$$

Thus we have

$$I_1 = \frac{E_s}{R_1 + Z_{11}(s)} \tag{7.8}$$

For $s = j\omega$, the power delivered to $Z_{11}(j\omega)$ is

$$P_1(j\omega) = |I_1|^2 R_{11}(\omega) = \frac{|E_s|^2 R_{11}(\omega)}{|R_1 + Z_{11}(j\omega)|^2} \tag{7.9}$$

Since the twoport is lossless, this power is also the power delivered to R_2, or $P_1 = P_2$. Hence

$$|t(j\omega)|^2 = \frac{\dfrac{|E_s|^2 R_{11}(\omega)}{|R_1 + Z_{11}(j\omega)|^2}}{\dfrac{|E_s|^2}{4R_1}} = \frac{4R_1 R_{11}(\omega)}{|R_1 + Z_{11}(j\omega)|^2} \tag{7.10}$$

Using (7.6),

$$|\rho(j\omega)|^2 = 1 - \frac{4R_1 R_{11}(\omega)}{[R_1 + R_{11}(\omega)]^2 + [X_{11}(\omega)]^2}$$

$$= \frac{[R_1 - R_{11}(\omega)]^2 + [X_{11}(\omega)]^2}{[R_1 + R_{11}(\omega)]^2 + [X_{11}(\omega)]^2} = \frac{|R_1 - Z_{11}(j\omega)|^2}{|R_1 + Z_{11}(j\omega)|^2} \tag{7.11}$$

Equation (7.11) will be satisfied if we make

$$\rho(s) = \pm \frac{R_1 - Z_{11}(s)}{R_1 + Z_{11}(s)} \tag{7.12}$$

Solving for $Z_{11}(s)$ from (7.12), we obtain

$$Z_{11}(s) = R_1 \frac{1 \pm \rho(s)}{1 \mp \rho(s)} \tag{7.13}$$

Thus the problem of realizing a given $|t(j\omega)|^2$ is replaced by one of realizing a driving-point impedance - $Z_{11}(s)$ - in the form of a lossless twoport terminated in a resistance.

Darlington [Da] has shown that any realizable driving-point impedance can be realized as a lossless twoport terminated in a resistance. The general proof of this assertion requires the use of unity-coupled mutual inductances or ideal transformers. Fortunately, many useful filter circuits can be realized with lossless twoports that do not use either mutual inductances or ideal transformers. Otherwise, the use of these devices can be avoided by a slight modification of the specified transmission coefficient.

In the arrangement of Fig. 7.1, routine analysis will give

$$\frac{E_2}{E_s} = \frac{z_{21}}{(z_{22} + R_2)(z_{11} + R_1) - z_{21}z_{12}} \tag{7.14}$$

From (7.4) and (7.14), we can infer that the zeros of z_{21} of the lossless twoport are the transmission zeros of $t(s)$. This is a very important observation. For example, if $t(s)$ is an all-pole function, then all transmission zeros will be at infinity. Then we can expect the lossless twoport to have the general configuration shown in Fig. 7.3(a). If all transmission zeros are at the origin, as in the case of a certain highpass filter, then we can expect the lossless twoport to have the general configuration of Fig. 7.3(b). If a pair of transmission zeros is present, we can expect either a series LC parallel combination or a shunt LC series branch as shown in Fig. 7.3(c) and (d).

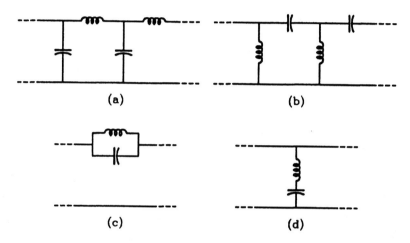

Figure 7.3: LC twoport configurations expected from transmission zeros.

7.2 LC ladders with equal terminations

We are now in a position to implement some of the relationships developed and observations made in the preceding section to realize filters in doubly-terminated form. We shall first deal with situations in which $R_1 = R_2$. We shall normalize both resistances to 1 ohm. When the two resistances are not equal, certain complications arise. We shall deal with those points in the next section.

We shall recapitulate the steps involved in the realization of doubly-terminated filters. We start off with a given $t(s)$ or $|t(j\omega)|^2$. From this we obtain $|\rho(j\omega)|^2$ using (7.6). From $|\rho(j\omega)|^2$, we can get $\rho(s)$. This can be done according to the procedure described in Section 3.1. Sometimes, certain parts of this function can be obtained by other considerations.

For example, the denominators of $t(s)$ and $\rho(s)$ are identical. Also, if we should find $|t(j\omega_0)| = 1$, then $\rho(s)$ will have zeros at $\pm j\omega_0$.

Once $\rho(s)$ is obtained, we use (7.13) to obtain $Z_{11}(s)$. Finally, we realize the driving-point impedance as a lossless twoport terminated in a resistance. We shall limit our developments to cases in which the lossless twoport is a ladder.

EXAMPLE 7.1. Assuming $R_1 = R_2 = 1$ Ω, realize the third-order Butterworth lowpass filter with

$$|t(j\omega)|^2 = \frac{1}{1 + \omega^6} \tag{7.15}$$

SOLUTION   We obtain

$$|\rho(j\omega)|^2 = \frac{\omega^6}{1 + \omega^6}$$

With the aid of Table A.2, we can write

$$\rho(s) = \frac{s^3}{s^3 + 2s^2 + 2s + 1}$$

Suppose we choose the upper signs of (7.13). We then have

$$Z_{11}(s) = \frac{1 + \rho(s)}{1 - \rho(s)} = \frac{2s^3 + 2s^2 + 2s + 1}{2s^2 + 2s + 1} \tag{7.16}$$

We can now apply Foster's preamble by continually removing poles at infinity. The process can be performed by the following long division.

$$
\begin{array}{r}
s\ \Omega \\
2s^2 + 2s + 1 \overline{\smash{\big)}\ 2s^3 + 2s^2 + 2s + 1} \\
\underline{2s^3 + 2s^2 + s} \qquad\qquad 2s\ \mho \\
s + 1 \overline{\smash{\big)}\ 2s^2 + 2s + 1} \\
\underline{2s^2 + 2s} \qquad\qquad s + 1\ \Omega \\
1 \overline{\smash{\big)}\ s + 1} \\
s + 1
\end{array}
$$

The resulting circuit is shown in Fig. 7.4.

The solution of this example can also be performed with MATLAB. The following are the steps involved.

```
≫ rho = 's^3/(s^3 + 2*s^2 + 2*s + 1)';
≫ num = symadd(1,rho);
```

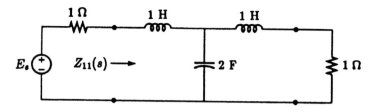

Figure 7.4: Circuit realizing (7.15).

```
» den = symsub(1,rho);
» z11 = simple(symdiv(num,den))
z11 = (2*s^3 + 2*s^2 + 2*s + 1)/(2*s^2 + 2*s + 1)
» z2 = simple(symsub(z11,'s'))
z2 = (s + 1)/(2*s^2 + 2*s + 1)
» y2 = symdiv(1,z2)
y2 = 1/(s + 1)*(2*s^2 + 2*s + 1)
» y3 = simple(symsub(y2,'2*s'))
y3 = 1/(s + 1)
```

If we had chosen the lower signs of (7.13) instead, we would have had

$$Z_{11}(s) = \frac{1 - \rho(s)}{1 + \rho(s)} = \frac{2s^2 + 2s + 1}{2s^3 + 2s^2 + 2s + 1} \qquad (7.17)$$

which is simply the reciprocal of the right-hand side of (7.16). The realization of this new $Z_{11}(s)$ is simply the dual of that for (7.16). The circuit of Fig. 7.5 would have been obtained. Both circuits obtained have exactly the same $|t(j\omega)|$.

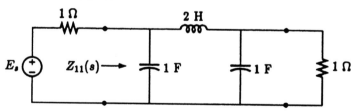

Figure 7.5: Another circuit realizing (7.15).

It's instructive to make a couple of observations borne out by this example. First, the fact that we can keep applying Foster's preamble steps to remove poles at infinity is because of the fact that we knew in advance that all transmission zeros are at infinity. The resulting LC ladders of the twoports are of the form of Fig. 7.3(a).

That the last element in the development of the ladders turned out to be 1-ohm resistances is not a coincidence. Since $t(0) = 1$ and the Butterworth characteristic is a monotonically decreasing function, the maximum of $|t(j\omega)|$ occurs at $\omega = 0$. At $s = 0$, all capacitors become open circuits, and all inductors become short circuits. So, at $s = 0$, the load resistance is connected directly to the Thévenin equivalent of the source. Hence to achieve $t(0) = 1$, we must have $R_2 = 1 \, \Omega$.

By a similar reasoning, if we are realizing a highpass filter whose transmission zeros are all at the origin, we can anticipate that $Z_{11}(s)$ can be realized by a ladder of the form of Fig. 7.3(b). Foster's preamble can again be applied to realize the ladder.

EXAMPLE 7.2. Realize a highpass filter with $R_1 = R_2 = 1 \, \Omega$ with equal-ripple variation in the pass band, $\alpha_p = 1$ dB, $n = 3$, and $\omega_p = 1$.

SOLUTION   For a normalized lowpass Chebyshev characteristic, we have

$$|t_{\mathrm{LP}}(j\omega)|^2 = \frac{1}{1 + 0.258925C_3^2(\omega)}$$

Using the lowpass-to-highpass transformation, we obtain

$$|t_{\mathrm{HP}}(j\omega)|^2 = \frac{1}{1 + 0.258925C_3^2\left(\dfrac{1}{\omega}\right)} \tag{7.18}$$

and

$$|\rho(j\omega)|^2 = \frac{0.258925C_3^2\left(\dfrac{1}{\omega}\right)}{1 + 0.258925C_3^2\left(\dfrac{1}{\omega}\right)} \tag{7.19}$$

To obtain the $\rho(s)$ for the magnitude-squared function of (7.19), we do not have to rely on the routine procedure outlined in Section 3.1. First, the denominator polynomial of $t_{\mathrm{LP}}(s)$ is given in Table A.8 as

$$Q_{\mathrm{LP}}(s) = s^3 + 0.988341s^2 + 1.238409s + 0.491307$$

To obtain the denominator polynomials of $t_{HP}(s)$, we simply reverse the powers of the s terms, resulting in

$$1 + 0.988341s + 1.238409s^2 + 0.491307s^3$$

For neatness, we make the coefficient of the cubic term unity. Thus,

$$Q_{HP}(s) = s^3 + 2.52062s^2 + 2.01165s + 2.03538$$

This is also the denominator of $\rho(s)$.

To obtain the numerator of $\rho(s)$, we observe that $|\rho(j\omega)|$ is zero when $|t_{HP}(j\omega)|$ is unity. We already know that $|t_{HP}(j\omega)|$ is unity at infinity. In addition, it is unity when $C_3(1/\omega) = 0$ or

$$\cos\left[3\cos^{-1}\left(\frac{1}{\omega}\right)\right] = 0 \quad \Longrightarrow \quad \omega = \pm\frac{2}{\sqrt{3}}$$

Hence we can construct

$$\rho(s) = \frac{K(s^2 + 4/3)}{s^3 + 2.52062s^2 + 2.01165s + 2.03538}$$

The only unknown number is the constant K. Since $t(0) = 0$, we must have $\rho(0) = 1$. Thus $K = 1.52653$ and

$$\rho(s) = \frac{1.52653(s^2 + 4/3)}{s^3 + 2.52062s^2 + 2.01165s + 2.03538}$$

To obtain $Z_{11}(s)$, we arbitrarily choose the upper signs of (7.13) to get

$$Z_{11}(s) = \frac{s^3 + 4.04715s^2 + 2.01165s + 4.07075}{s^3 + 0.994089s^2 + 2.01165s}$$

Note that $Z_{11}(\infty) = 1$. We next perform the following long division, similar to the Cauer 2 steps, by keeping removing poles at the origin.

$$
\begin{array}{r}
\frac{1}{0.49417}/s \ \Omega \\
2.01165s + 0.994089s^2 + s^3 \overline{\smash{\big)}\ 4.07075 + 2.01165s + 4.04715s^2 + s^3} \\
4.07075 + 2.01165s + 2.02355s^2 \\
\hline
2.02355s^2 + s^3
\end{array}
$$

$$\frac{\frac{1}{1.00593}/s \; \mho}{2.01165s + 0.994089s^2 + s^3}$$

$$2.02355s^2 + s^3 \overline{\left| \begin{array}{c} 2.01165s + 0.994089s^2 + s^3 \\ 2.01165s + 0.994089s^2 \end{array} \right.}$$

$$\frac{1}{0.49417}/s + 1 \; \Omega$$

$$s^3 \overline{\left| \begin{array}{c} 2.02355s^2 + s^3 \\ 2.02355s^2 + s^3 \end{array} \right.}$$

The resulting circuit is shown in Fig. 7.6.

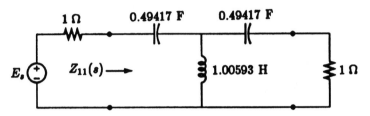

Figure 7.6: Circuit that realizes the highpass transmission characteristic of (7.18).

We could have worked Example 2 by first obtaining a lowpass filter with the same specification and then applying lowpass-to-highpass transformation. The result would have been exactly the same. We chose to obtain the highpass filter directly to illustrate the application of Foster's preamble to this type of realization.

Also, there is another circuit that will realize the same filtering characteristic if we use the lower signs of (7.13). The resulting circuit is the dual of the circuit of Fig. 7.6.

If transmission zeros are required on the finite nonzero points on the $j\omega$ axis, zero shifting can also be applied to $Z_{11}(s)$ similar to when the technique is used on lossless functions. The following example will illustrate this situation.

EXAMPLE 7.3. The following transmission coefficient realizes a third-order elliptic filter with $\alpha_p = 1$ dB, $\omega_s = 1.5$, $\omega_p = 1$, and $\alpha_s = 25.17$ dB.

$$t(s) = \frac{0.215619(s^2 + 2.80601)}{(s + 0.591015)(s^2 + 0.375396s + 1.02371)} \qquad (7.20)$$

Obtain the filter circuit in the form of doubly-terminated LC ladder with $R_1 = R_2 = 1 \; \Omega$.

SOLUTION The following steps are straightforward.

$$|t(j\omega)|^2 = \frac{0.0464916\omega^4 - 0.260912\omega^2 + 0.366062}{\omega^6 - 1.55721\omega^4 + 0.382050\omega^2 + 0.366062} \qquad (7.21)$$

$$|\rho(j\omega)|^2 = \frac{\omega^6 - 1.60370\omega^4 + 0.642962\omega^2}{\omega^6 - 1.55721\omega^4 + 0.382050\omega^2 + 0.366062} \qquad (7.22)$$

$$\rho(s)\rho(-s) = \frac{s^6 + 1.60370s^4 + 0.642962s^2}{s^6 + 1.55721s^4 + 0.382050s^2 - 0.366062} \qquad (7.23)$$

$$\rho(s) = \frac{s(s^2 + 0.801849)}{(s + 0.591015)(s^2 + 0.375396s + 1.02371)} \qquad (7.24)$$

Arbitrarily, we choose the lower signs of (7.13) to get

$$Z_{11}(s) = \frac{0.966411s^2 + 0.443730s + 0.605031}{2s^3 + 0.966411s^2 + 2.04743s + 0.605031} \qquad (7.25)$$

Now we remove part of the pole at infinity from $1/Z_{11}$ to produce a pair of zeros at $\pm j\sqrt{2.80601}$. This is done by removing a shunt capacitor of C_1 farads. We require

$$\frac{1}{Z_{11}(s)} - C_1 s = 0 \qquad \text{for} \qquad s = j\sqrt{2.80601} \qquad (7.26)$$

Solving yields $C_1 = 1.69200$ F. The remainder is

$$y_2 = \frac{1}{Z_{11}(s)} - 1.69200s = \frac{s^3 + 0.591015s^2 + 2.80601s + 1.65840}{2.64895s^2 + 1.21627s + 1.65840} \qquad (7.27)$$

The numerator of y_2 must have a factor $(s^2 + 2.80601)$. Hence $1/y_2$ must have a pair of poles at $\pm j\sqrt{2.80601}$. Some algebra work will give

$$z_2 = \frac{1}{y_2} = \frac{2.64895s^2 + 1.21627s + 1.65840}{(s^2 + 2.80601)(s + 0.591053)} \qquad (7.28)$$

We shall next remove this pair of poles. The terms to be removed can be written as $k_1 s/(s^2 + 2.80601)$. Using (5.8), we find

$$k_1 = \left[\frac{2.64895s^2 + 1.21627s + 1.65840}{s(s + 0.591015)} \right]_{s=j\sqrt{2.80601}} = 2.05793 \qquad (7.29)$$

The remainder is

$$z_3 = z_2 - \frac{2.05793s}{s^2 + 2.80601} = \frac{0.591015}{s + 0.591015} \qquad (7.30)$$

$$y_3 = \frac{1}{z_3} = 1.69200s + 1 \qquad (7.31)$$

The resultant network is shown in Fig. 7.7.

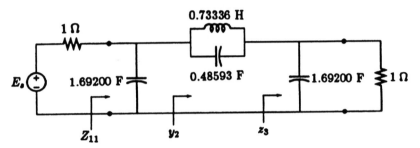

Figure 7.7: Circuit realizing the transmission coefficient of (7.20).

The following are the steps used to solve the same example using MATLAB. Starting with (7.23), we have the coefficient vectors of the numerator and denominator polynomials of $\rho(s)\rho(-s)$. We first obtain the $\rho(s)$ of (7.24).

```
≫ nrho2 = [1 0 1.60370 0 0.642962 0 0];
≫ n = roots(nrho2)
n = 0; 0; + 0.8955i; - 0.8955i; + 0.8954i; -0.8954i
≫ nrho = poly([n(1) n(3) n(4)])
nrho = 1 0 0.8019 0
≫ drho2 = [1 0 1.55721 0 0.382050 0 -0.366062];
≫ d = roots(drho2)
d = -0.1877 + 0.9942i; -0.1877 - 0.9942i; 0.1877 + 0.9942i;
0.1877 - 0.9942i; 0.5910; -0.5910
```

The following is the quadratic factor in the denominator of (7.24).

```
≫ f1 = poly([d(1) d(2)])
f1 = 1 0.3754 1.0237
```

The following is the denominator of $\rho(s)$ of (7.24) in expanded form.

```
» drho = conv(f1,[1,-d(6)])
drho = 1 0.9664 1.2456 0.6050
```

The following gives the numerator and denominator of Z_{11} in (7.25).

```
» nz11 = drho-nrho
nz11 = 0 0.9664 0.4436 0.6050
» dz11 = drho+nrho
dz11 = 2.0000 0.9664 2.0475 0.6050
```

From (7.26), we have

$$C_1 = \left[\frac{1}{sZ_{11}(s)}\right]_{s=j\sqrt{2.80601}} = \left[\frac{\text{Den}[Z_{11}]}{s\text{Num}[Z_{11}]}\right]_{s=j\sqrt{2.80601}}$$

We first augment the numerator (of Z_{11}) vector with a zero. This is the result of multiplying the numerator by s.

```
» nz11a = [nz11 0]
nz11a = 0 0.9664 0.4436 0.6050 0
» C1 = freqs(dz11,nz11a,sqrt(2.80601))
C1 = 1.6920
```

To obtain y_2, we realize that

$$y_2 = \frac{\text{Den}[Z_{11}] - C_1 s\text{Num}[Z_{11}]}{\text{Num}[Z_{11}]}$$

To obtain the second term of the numerator, we remove the leading zero of nz11 and augment the vector with another zero.

```
» nz11a = [nz11(1,2:4) 0]
nz11a = 0.9664 0.4436 0.6050 0
» ny2 = dz11-c1*nz11a
ny2 = 0.3648 0.2158 1.0238 0.6050
» dy2 = nz11(1,2:4)
dy2 = 0.9664 0.4436 0.6050
```

To make the leading coefficient of the numerator of y_2 unity we divide
ny2 and dy2 by ny2(1).

```
» c = ny2(1);
» ny2 = ny2/c
ny2 = 1 0.5910 2.8060 1.6584
» dy2 = dy2/c
dy2 = 2.6489 1.2163 1.6584
```

which is the same as what is given in (7.27). To find the poles of z_2, we
determine the zeros of y_2.

```
» pz2 = roots(ny2)
pz2 = + 1.6751i; - 1.6751i; -0.5911
```

which confirms the pair of poles at $\pm j\sqrt{2.80601}$. The following implements (7.29).

```
» k1 = freqs(dy2,[1 -pz2(3) 0],sqrt(2.80601))
k1 = 2.0579
```

To obtain z_3, we find the partial fraction expansion of $z_2 = 1/y_2$.

```
» [r,p,k] = residue(dy2,ny2)
r = 1.0290; 1.0289; 0.5911
p = + 1.6751i; - 1.6751i; -0.5910
k = []
```

Combining the pair of conjugate poles, we get

$$z_2 = \frac{2.0580s}{s^2 + 2.8060} + \frac{0.5911}{s + 0.5910}$$

which, allowing for truncation errors, agrees well with (7.30). The last
step, (7.31), is obvious.

7.3   LC ladders with unequal terminations

In the preceding section, we have limited our examples to doubly-
terminated LC ladders with equal terminations - $R_1 = R_2$. If $R_1 \neq R_2$,
some difficulties will arise in the realization of some filters. For example,

suppose $R_1 = 1\,\Omega$ and $R_2 = 4\,\Omega$ and we wish to realize the $|t(j\omega)|^2$ of (7.15). This problem has already been worked out in Example 7.1 in which two circuits were obtained. But both circuits require the terminating resistances to be 1 ohm at both ends of the networks.

One simple solution would be to replace the 1-ohm resistance (R_2) by the parallel combination of a 4-ohm resistor and a 4/3-ohm one as shown in Fig. 7.8. We can then consider the 4/3-ohm resistance as part of the insertion network. However, this will mean a four-to-one reduction (nominally 6 dB) in power delivered to the actual load - the 4-ohm resistor.

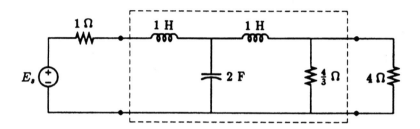

Figure 7.8: One method of accommodating unequal terminations.

Another alternative is to use an ideal transformer as shown in Fig. 7.9. The 4-ohm resistance appears as a 1-ohm one from the left-hand side of the ideal transformer. Since the ideal transformer is lossless, the transmission coefficient in unchanged.

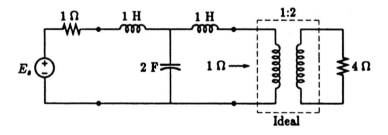

Figure 7.9: Using an ideal transformer to accommodate unequal terminations.

Another possible recourse is to use the original Darlington synthesis. However, this method not only results in a very elaborate network, but also necessitates the use of closely coupled mutual inductances and, in some cases, ideal transformers.

The most practical method of accommodating unequal terminations is to anticipate the reduction in gain due to the less than maximum power transfer when the two terminating resistances are connected directly to each other. This method is best explained if we make an observation on the performance of the lowpass ladder filters at low frequencies. For $s = 0$, all inductances become short circuits and all capacitances become open circuits. We can usually expect a lowpass ladder to degenerate into a direct connection of R_1 and R_2 as shown in Fig. 7.10.

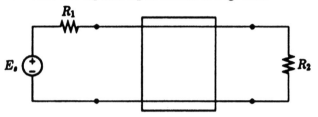

Figure 7.10: Circuit of a doubly-terminated LC ladder at $s = 0$.

In Fig. 7.10, we have

$$t^2(0) = \frac{\dfrac{|E_s|^2 R_2}{(R_1 + R_2)^2}}{\dfrac{|E_s|^2}{4R_1}} = \frac{4R_1 R_2}{(R_1 + R_2)^2} \leq 1 \tag{7.32}$$

In (7.32), if $R_1 = R_2$, then $t(0) = 1$. If $R_1 \neq R_2$, then $t(0) < 1$. So, if $R_1 \neq R_2$ and if we accept this slight reduction in gain by making $t^2(0)$ equal to what (7.32) gives, we can anticipate that the desired R_2 will result as $Z_{11}(s)$ is realized by Foster's preamble.[1]

EXAMPLE 7.4. Realize a third-order Butterworth lowpass filter for $R_1 = 1\ \Omega$ and $R_2 = 4\ \Omega$.

SOLUTION   From (7.32),

$$t^2(0) = \frac{4 \times 1 \times 4}{(1 + 4)^2} = 0.64$$

We let

$$|t(j\omega)|^2 = \frac{0.64}{1 + \omega^6} \tag{7.33}$$

---

[1]As we shall see soon, sometimes we may end up with $1/R_2$ instead of R_2.

This represents a 1.94 dB reduction in gain from what is given in (7.15). We now have

$$|\rho(j\omega)|^2 = \frac{\omega^6 + 0.36}{\omega^6 + 1} \tag{7.34}$$

We can make

$$\rho(s) = \frac{s^3 + 0.6}{s^3 + 2s^2 + 2s + 1}$$

and choose

$$Z_{11} = \frac{1 + \rho(s)}{1 - \rho(s)} = \frac{2s^3 + 2s^2 + 2s + 1.6}{2s^2 + 2s + 0.4}$$

Applying Foster's preamble results in the following long division.

$$
\begin{array}{r}
s\ \Omega \\
2s^2 + 2s + 0.4 \enclose{longdiv}{2s^3 + 2s^2 + 2s + 1.6} \\
\underline{2s^3 + 2s^2 + 0.4s} \qquad \tfrac{5}{4}s\ \mho \\
1.6s + 1.6 \enclose{longdiv}{2s^2 + 2s + 0.4} \\
\underline{2s^2 + 2s} \qquad 4s + 4\ \Omega \\
0.4 \enclose{longdiv}{1.6s + 1.6} \\
\underline{1.6s + 1.6}
\end{array}
$$

Gathering the results of the long division, we obtain the circuit of Fig. 7.11.

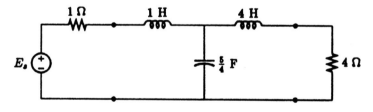

Figure 7.11: Circuit that realizes (7.33).

In Example 7.4, we obtained only one of several possible circuits. One of the steps that has multiple answers involves obtaining $\rho(s)$ from $|\rho(j\omega)|^2$. If we let $\rho(s) = P(s)P(-s)$, then for $|\rho(j\omega)|^2$ in (7.34) we have

$$P(s)P(-s) = s^6 - 0.36$$

The right-hand side of (7.34) has six zeros equally spaced along a circle of radius $\sqrt[6]{0.36}$. Since $P(s)$ does not have to be Hurwitz, there are four distinct possible combinations for $P(s)$. They are

$$P(s) = s^3 + 0.6 \tag{7.35}$$

$$P(s) = s^3 - 0.6 \tag{7.36}$$

$$P(s) = s^3 + 1.68687s^2 + 1.42276s + 0.6 \tag{7.37}$$

$$P(s) = s^3 - 1.68687s^2 + 1.42276s - 0.6 \tag{7.38}$$

In Example 7.4, we used the $P(s)$ of (7.35). If we had used the $P(s)$ of (7.37), we would have obtained

$$Z_{11}(s) = \frac{2s^3 + 3.68687s^2 + 3.42276s + 1.6}{0.313135s^2 + 0.577243s + 0.4} \tag{7.39}$$

Applying Foster's preamble results in the circuit of Fig. 7.12.

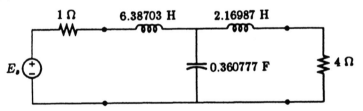

Figure 7.12: Another circuit that realizes (7.33).

If we had used the $P(s)$ of (7.38) and still used the upper signs of (7.13), we would have obtained

$$Z_{11}(s) = \frac{2s^3 + 0.313135s^2 + 3.42276s + 0.4}{3.68687s^2 + 0.577243s + 1.6} \tag{7.40}$$

The circuit of Fig. 7.13 would result.

The last resistance obtained for $Z_{11}(s)$ in (7.40) is $\frac{1}{4}$ ohm instead of 4 ohms. The reason why this occurs is that

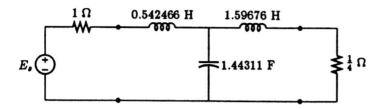

Figure 7.13: Another circuit that realizes (7.33).

$$t^2(0) = \frac{4R_1R_2}{(R_1 + R_2)^2} = \frac{4}{\left(\sqrt{\dfrac{R_1}{R_2}} + \sqrt{\dfrac{R_2}{R_1}}\right)^2} \tag{7.41}$$

is a symmetric function with respect to R_1 and R_2. Therefore, whether $R_2/R_1 = 4$ or $R_1/R_2 = 4$, the same $t(0)$ results. Hence, making $t^2(0)$ satisfy (7.32) merely leads to a certain resistance ratio, but not necessarily in any particular order.

One alternative is to use the reciprocal of (7.40). This amounts to choosing the lower signs of (7.13). The development of the reciprocal of (7.40) as an impedance function will lead to the circuit of Fig. 7.14.

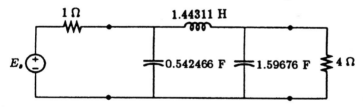

Figure 7.14: Circuit obtained by developing the reciprocal of (7.40) as an impedance.

Another alternative to accommodate the 1-ohm source resistance and the 4-ohm load resistance using the circuit of Fig. 7.13 is to reverse the input and the output and then apply an impedance scaling. This alternative will be explained in the next section.

7.4   A doubly-terminated filter used in reverse

If an LC network has been obtained for a given $|t(j\omega)|^2$, from Fig. 7.1 and (7.4), we have

$$|t(j\omega)|^2 = \frac{4R_1}{R_2}\left|\frac{E_2}{E_s}\right|^2 = 4R_1R_2\left|\frac{I_2}{E_s}\right|^2 \qquad (7.42)$$

If we apply the reciprocity theorem to obtain the transmission in the opposite direction, we have the arrangement of Fig. 7.15. Reciprocity requires that

$$\frac{I_1'}{E_s'} = \frac{I_2}{E_s} \qquad (7.43)$$

The transmission from E_s' to R_1 is

$$|t'(j\omega)|^2 = \frac{|I_1'|^2 R_1}{\dfrac{|E_s'|}{4R_2}} = 4R_1R_2\left|\frac{I_1'}{E_s'}\right|^2 = 4R_1R_2\left|\frac{I_2}{E_s}\right|^2 = |t(j\omega)|^2 \quad (7.44)$$

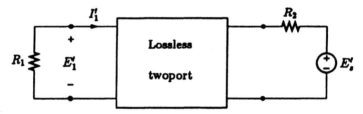

Figure 7.15: A doubly-terminated filter used in reverse.

Hence the transmissions in both directions of a doubly-terminated lossless twoport are identical. Furthermore, since

$$|t(j\omega)|^2 = \frac{4R_1}{R_2}\left|\frac{E_2}{E_s}\right|^2 = |t'(j\omega)|^2 = \frac{4R_2}{R_1}\left|\frac{E_1'}{E_s'}\right|^2 \qquad (7.45)$$

a doubly-terminated arrangement can be impedance-scaled without affecting the transmission coefficient.

As an example of the application of this idea, we can take the circuit of Fig. 7.13 and interchange the input and output as shown in Fig. 7.16. Then we impedance-scale the network by $k_z = 4$. The circuit of Fig. 7.17 results. It is seen that this circuit is identical to that shown in Fig. 7.12. Hence, the specification given in Example 7.4 can also be satisfied this way.

EXAMPLE 7.5. Realize a doubly-terminated LC ladder for $R_1 = 1\ \Omega$, $R_2 = 2\ \Omega$, with a Chebyshev lowpass characteristic in which $\alpha_p = 0.5$ dB and $n = 2$.

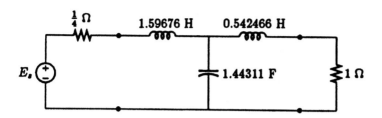

Figure 7.16: The circuit of Fig. 7.13 with its input and output reversed.

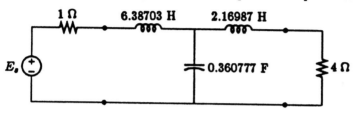

Figure 7.17: Circuit of Fig. 7.16 impedance-scaled by $k_z = 4$.

SOLUTION   We have $\epsilon^2 = 0.122018$ and

$$|t(j\omega)|^2 = \frac{K}{1 + 0.122018 C_2^2(\omega)} \tag{7.46}$$

We wish to have

$$t^2(0) = \frac{4R_1 R_2}{(R_1 + R_2)^2} = \frac{8}{9}$$

or

$$\frac{K}{1.122018} = \frac{8}{9} \quad \Longrightarrow \quad K = 0.997350$$

$$|\rho(j\omega)|^2 = 1 - \frac{0.997350}{1 + 0.122018(2\omega^2 - 1)^2} = \frac{\omega^4 - \omega^2 + 0.255429}{\omega^4 - \omega^2 + 2.29887}$$

$$\rho(s) = \frac{s^2 + 0.103922s + 0.505400}{s^2 + 1.42562s + 1.51620}$$

If we choose the upper signs of (7.13), we have

$$Z_{11}(s) = \frac{1 + \rho(s)}{1 - \rho(s)} = \frac{2s^2 + 1.52955s + 2.02160}{1.32170s + 1.01080} \tag{7.47}$$

After developing $Z_{11}(s)$, we obtain the circuit of Fig. 7.18.

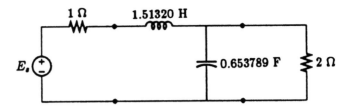

Figure 7.18: Circuit for Example 7.5.

If we chose the lower signs of (7.13), we would obtain the reciprocal of the function of (7.47). The circuit obtained would be the one shown in Fig. 7.19. We next turn the circuit end over end and impedance-scale it

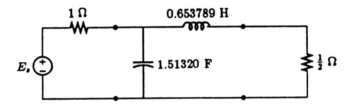

Figure 7.19: Another circuit for Example 7.5.

by $k_z = 2$. The circuit of Fig. 7.20 results.[2]

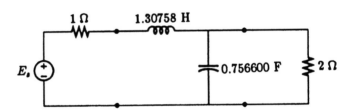

Figure 7.20: The circuit of Fig. 7.19 turned end over end and impedance-scaled by $k_z = 2$.

EXAMPLE 7.6. Determine whether or not it is possible to realize, without transformers, a doubly-terminated LC ladder Chebyshev low-pass filter with $R_1/R_2 = 3$, $\alpha_p = 2$ dB, and $n = 4$.

---

[2]This circuit is identical to the one that would have been obtained if we had let the numerator of $\rho(s)$ be $s^2 - 0.103922s + 0.505400$ and chosen the lower signs of (7.13).

SOLUTION

$$t^2(0) = \frac{4 \times 1 \times 3}{(1+3)^2} = \frac{3}{4}$$

$$|t(j\omega)|^2_{\max} = \frac{3}{4} \times 10^{0.2} = 1.1887 > 1$$

Therefore, the specified filter cannot be realized.

7.5 Summary

In this chapter, we have presented the method of realizing a ladder filter that is to be inserted between a source with an internal resistance and a resistive load. The source resistance usually represents the Thévenin's equivalent impedance of the circuit before the input of the filter. The load resistance represents the input impedance of the circuit or device that is connected to the output of the filter. In practice, this situation occurs far more frequently than the singly-terminated configurations treated in the previous chapter.

Problems

7.1 Obtain a doubly-terminated LC ladder with $R_1 = R_2 = 1\ \Omega$ to realize each of the following $|t(j\omega)|^2$.

(a) $\dfrac{\omega^4}{1+\omega^4}$

(b) $\dfrac{1}{(1+\omega^2)^2}$

(c) $\dfrac{1}{(1+\omega^2)^3}$

(d) $\dfrac{1}{1+\omega^8}$

(e) $\dfrac{1}{1+\omega^2(1-\omega^2)^2}$

7.2 Obtain a third-order Chebyshev doubly-terminated lowpass filter with $\alpha_p = 1$ dB, $R_1 = R_2 = 50\ \Omega$, and $\omega_p = 1000$ rad/sec.

7.3 Given

$$t(s) = \frac{Ks^3}{(s+1)^3}$$

determine the maximum value of K such that $t(s)$ is realizable as an LC ladder terminated in 1 ohm at both ends. Obtain the LC ladder.

7.4 Given that $R_1 = 1\ \Omega$, R_2 is arbitrary, and

$$|t(j\omega)|^2 = \frac{0.25}{\omega^4 - \omega^2 + 0.5}$$

obtain at least two networks to realize this transmission function.

7.5 Refer to Fig. 7.1. Determine all possible $Z_{11}(s)$ such that the following function will be realized.

$$|t(j\omega)|^2 = \frac{\omega^2}{\omega^4 + 2\omega^2 + 1}$$

7.6 We have a doubly-terminated LC ladder with $R_1 = 1\ \Omega$. A fourth-order lowpass Chebyshev filter is to be realized such that $\alpha_p = 1$ dB. What is the value of R_2 that will make the power delivered to R_2 in the pass band as high as possible?

7.7 Suppose $R_1 = 4\ \Omega$, $R_2 = 1\ \Omega$, and

$$t(s) = \frac{Ks}{s^2 + s + 1}$$

If ideal transformers are allowed, what is the maximum value of K? Obtain an insertion network to realize this $t(s)$.

7.8 It is necessary to insert a lowpass LC ladder between two resistances $R_1 = 1\ \Omega$ and $R_2 = 8\ \Omega$. The lowpass characteristic is to have equal ripples in the pass band $\omega_p = 1$ with $\alpha_p = 1$ dB and $n = 2$. Obtain the LC ladder.

7.9 Repeat the previous problem for $n = 3$.

7.10 Specify the appropriate $|t(j\omega)|^2$ for a fourth-order Chebyshev lowpass filter without transformers to work between $R_1 = 6\ \Omega$ and $R_2 = 1\ \Omega$ with $\alpha_p = 1$ dB.

7.11 For $R_1 = 1 \ \Omega$ and

$$|t(j\omega)|^2 = \frac{K}{1 + 0.25C_2^2(\omega)}$$

determine the value of K such that $|t(j\omega)|_{\max} = 1$. If, in addition, the insertion network is to be transformerless, what is the appropriate value of R_2?

7.12 Obtain an LC ladder to be inserted between $R_1 = 1 \ \Omega$ and $R_2 = 3 \ \Omega$ and

$$|t(j\omega)|^2 = \frac{K}{1 + \omega^4}$$

7.13 Realize an LC ladder to be inserted between $R_1 = 1 \ \Omega$ and $R_2 = 5 \ \Omega$ to give a second-order lowpass Chebyshev filter with $\alpha_p = 2$ dB.

7.14 Obtain an LC ladder to be inserted between $R_1 = 1 \ \Omega$ and $R_2 = 5 \ \Omega$ and

$$|t(j\omega)|^2 = \frac{K}{1 + \omega^6}$$

7.15 Design a sixth-order bandpass filter based on a third-order Chebyshev lowpass prototype such that $\alpha_p = 1$ dB and the pass band is between 9 kHz and 11 kHz. Choose the zeros of $\rho(s)$ to be all in the left halfplane. The source and load resistances are 75 ohm and 300 ohm respectively.

7.16 For $R_1 = 1 \ \Omega$ and $R_2 = 5 \ \Omega$, find an LC ladder to realize a fourth-order Chebyshev lowpass transmission coefficient with $\alpha_p = 0.5$ dB.

7.17 Repeat the previous problem for $n = 5$.

7.18 For $R_1 = R_2 = 1 \ \Omega$, obtain a lossless ladder to realize the transmission coefficient

$$t(s) = \frac{2(s^2 + 1)}{3s^3 + 4s^2 + 3s + 2}$$

7.19 For $R_1 = 1 \ \Omega$ and $R_2 = \frac{1}{2} \ \Omega$, obtain a lossless ladder to realize

$$|t(j\omega)|^2 = \frac{8(\omega^2 - 2)^2}{49\omega^6 + 29\omega^4 - 56\omega^2 + 36}$$

7.20 An even-ordered elliptic-function filter is not realizable by a doubly-terminated LC ladder. Explain why by investigating the case when $n = 4$ and $t(s)$ has the form

$$t(s) = \frac{K(s^2 + \omega_1^2)(s^2 + \omega_2^2)}{s^4 + b_3 s^3 + b_2 s^2 + a_1 s + a_0}$$

Chapter 8

Sensitivity

The problem of finding a network to realize a certain network function is an open-ended one. There is no limit to how many networks one may be able to find to realize a given network function. This feature is especially true when active networks are used as we shall study in the upcoming chapters. In selecting a particular network from a number of available ones, different criteria may be used to make the final decision. Usually, the selection is based on the economics of the construction of the filter circuit. For example, in passive filters, the determining factor may be the number of inductors or the total inductance necessary. This is because inductors are usually the most expensive type of components, they deviate most from the idealized model, and they are most difficult to calibrate.

Sometimes, even economic factors are not the most important consideration. The final decision may be entirely subjective. For example, what size and type of components happen to be in stock will dictate which network to choose. The expertise of the hardware technicians of different companies may determine their preference in the selection of the networks adopted. Whether an active or a passive filter is to be used may depend on whether a power supply is readily available or not.

Another factor that may determine which of the circuits to use is the *sensitivity*. When we synthesize a number of networks to realize the same network function, we obtain their *nominal* component values. If all components are not only accurate, but also remain constant throughout their lifetime and under different ambient conditions, then there will be little difference in their performance. However, in practice, real components do deviate from their nominal values due to initial inaccuracy in fabrication, environmental factors, such as temperature and humidity, and the chemical and mechanical changes due to aging. As a result, component values may deviate from their nominal values. The effects of these deviations will alter the performance of different networks by different amounts.

There are two different approaches to reduce these undesirable effects. One is to make the initial component values very accurate and subsequently endeavor to keep these values from fluctuating. This approach usually is not only unnecessarily costly, but also impractical. The other approach is to select networks whose performances are least affected by the changes in their element values. This second approach is somewhat tractable mathematically by comparing the *sensitivities* of various network performance measures with respect to different component value changes.

For example, the two circuits in Fig. 8.1 both realize the transfer function

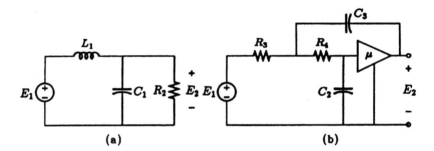

(a)   (b)

Figure 8.1: Two circuit realizing the voltage ratio of (8.1).

$$H(s) = \frac{E_2}{E_1} = \frac{1.10251}{s^2 + 1.09773s + 1.10251} \qquad (8.1)$$

if $L_1 = 0.99566$ H, $C_1 = 0.91097$ F, $R_2 = 1\ \Omega$, $\mu = 1$, $R_3 = 1.01430\ \Omega$, $R_4 = 8.94222\ \Omega$, $C_2 = 0.1$ F, and $C_3 = 1$ F. They both realize the second-order Chebyshev lowpass characteristic with $\alpha_p = 1$ dB as shown in Fig. 8.2. At $\omega = 1$, $|H(j\omega)| = 1$.

Now suppose, for some reason, the passive element values are changed slightly. Let's assume that all passive element values are increased by 1%.[1] Direct calculations will give

$$|H(j)| = 0.99168$$

---

[1] R_2 is not varied because it represents the load resistances.

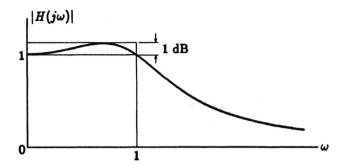

Figure 8.2: Magnitude characteristic for the circuits of Fig. 8.1.

for the circuit of Fig. 8.1(a). Or the gain at $\omega = 1$ is decreased by approximately 0.83%. If we make a similar assumption on all four passive elements in Fig. 8.1(b), we get

$$|H(j)| = 0.98308$$

In other words, the gain at $\omega = 1$ is decreased by approximately 1.7%. Hence, we may conclude that the active filter is approximately twice as sensitive as the passive circuit with respect to passive element value change.

The subject of sensitivity is a well-developed one in itself. We shall rely primarily on the simplest definition that will give us some indication of the relative merits of various filters as far as sensitivity is concerned. A detailed and comprehensive treatment of this subject is clearly beyond the scope of this text.

8.1 Definition of sensitivity

We shall define the sensitivity of some performance measure y with respect to a network element value x to be

$$S_x^y = \frac{x}{y} \cdot \frac{dy}{dx} \tag{8.2}$$

If y is a function of several variables $[y = f(x_1, x_2, \ldots, x_n)]$, then the sensitivity of y with respect to x_i is

$$S_{x_i}^y = \frac{x_i}{y} \cdot \frac{\partial y}{\partial x_i} \qquad i = 1, 2, \ldots, n \tag{8.3}$$

The significance of the definition in (8.2) is that for a small change in x, it gives approximately the ratio of the per-unit (or percent) change in y to the per-unit (or percent) change in x. In other words

$$S_x^y \approx \frac{\frac{\Delta y}{y}}{\frac{\Delta x}{x}} \tag{8.4}$$

Thus, roughly speaking, if $S = 2$, then a 1% change in x results in a 2% change in y. If $S = 0.1$, then a 1% change in x results in only a 0.1% change in y. Of course, these observations are only approximate. That is, (8.4) is only meaningful if $\Delta x/x$ is very small.

The definition given in (8.2) is known as the *differential, first-order, classical, relative,* or *Bode*[2] [Bo] sensitivity.

8.2 Properties of first-order sensitivity

Using the definition of (8.2), we may derive several useful identities of the first-order sensitivity.

(1) $\quad S_x^{ky} = \frac{x}{ky} \cdot \frac{d(ky)}{dx} = \frac{x}{y} \cdot \frac{dy}{dx} = S_x^y \tag{8.5}$

where k is independent of x.

(2) $\quad S_x^{y+k} = \frac{x}{y+k} \cdot \frac{d(y+k)}{dx} = \frac{y}{y+k} \cdot \frac{x}{y} \cdot \frac{dy}{dx} = \frac{y}{y+k} S_x^y \tag{8.6}$

where k is independent of x.

(3) $\quad S_{\frac{1}{x}}^y = \frac{\frac{1}{x}}{y} \cdot \frac{dy}{d\frac{1}{x}} = \frac{\frac{1}{x}}{y} \cdot \frac{dy}{dx} \cdot \frac{dx}{d\frac{1}{x}} = -\frac{x}{y} \cdot \frac{dy}{dx} = -S_x^y \tag{8.7}$

(4) $\quad S_x^{\frac{1}{y}} = \frac{x}{\frac{1}{y}} \cdot \frac{d\frac{1}{y}}{dx} = \frac{dy}{dx} = -S_x^y \tag{8.8}$

---

[2]Actually, the original definition of sensitivity given by Bode is the reciprocal of what is defined here.

(5) $\quad S_x^{y_1 y_2} = \dfrac{x}{y_1 y_2} \cdot \dfrac{d(y_1 y_2)}{dx} = \dfrac{x}{y_1 y_2} \cdot \left(y_1 \dfrac{dy_2}{dx} + y_2 \dfrac{dy_1}{dx} \right)$

$$= \dfrac{x}{y_1} \cdot \dfrac{dy_1}{dx} + \dfrac{x}{y_2} \cdot \dfrac{dy_2}{dx} = S_x^{y_1} + S_x^{y_2} \qquad (8.9)$$

(6) Combining (8.8) and (8.9), we have

$$S_x^{\frac{y_1}{y_2}} = S_x^{y_1} - S_x^{y_2} \qquad (8.10)$$

The following properties are listed without proof. The reader should try to complete the proofs as exercises.

(7) $\quad S_{x^n}^{y} = \dfrac{1}{n} S_x^{y}$ $\qquad\qquad\qquad\qquad\qquad\quad$ (8.11)

(8) $\quad S_x^{y^n} = n S_x^{y}$ $\qquad\qquad\qquad\qquad\qquad\quad\ $ (8.12)

(9) $\quad S_x^{y} = S_z^{y} - S_x^{z}$ $\qquad\qquad\qquad\qquad\qquad\ $ (8.13)

(10) $\quad S_x^{y} = S_x^{|y|} + j(\arg y) S_x^{(\arg y)}$ $\qquad\qquad\ $ (8.14)

(11) $\quad S_x^{y_1 + y_2} = \dfrac{y_1 S_x^{y_1} + y_2 S_x^{y_2}}{y_1 + y_2}$ $\qquad\qquad\ $ (8.15)

(12) $\quad S_x^{\ln y} = \dfrac{1}{\ln y} \cdot S_x^{y}$ $\qquad\qquad\qquad\qquad\ $ (8.16)

(13) $\quad S_x^{e^y} = y S_x^{y}$ $\qquad\qquad\qquad\qquad\qquad\quad\ $ (8.17)

8.3   Sensitivities of network performance

In assessing the sensitivities of the network performance with respect to an element value change, one of the most obvious parameters is the gain at a certain frequency, frequently near the band edge. This was done for the example circuits of Fig. 8.1 and we directly calculated the effect of $|H(j)|$ when C's and R's are changed. These changes are directly related to the sensitivity $S_x^{|H(j)|}$; $x = L_1, C_1, C_2, C_3, R_2, R_3, R_4$.

There are several other performance measures that can also be very informative about the sensitivity of the performance of a network due to some element value change. For a second-order bandpass filter with the network function

$$H(s) = \frac{Ks}{s^2 + b_1 s + b_0} \tag{8.18}$$

We could write the expression in the following form:

$$H(s) = \frac{Ks}{s^2 + \left(\dfrac{\omega_0}{Q}\right)s + \omega_0^2} \tag{8.19}$$

in which $Q = \sqrt{b_0}/b_1$ and $\omega_0 = \sqrt{b_0}$. The quantity ω_0 represents the center frequency of the magnitude characteristic as shown in Fig. 8.3(a). Hence the sensitivity $S_x^{\omega_0}$ gives an indication of the relative amount of $\Delta\omega_0$ with respect to an element change. The quantity Q represents the sharpness of the bandpass response and is related to the half-power bandwidth, denoted by BW, by the relationship

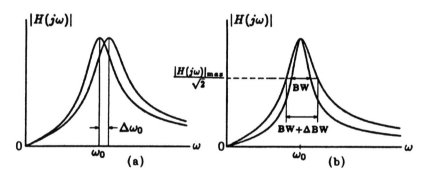

Figure 8.3: Effects of changes in Q and ω_0 on the bandpass characteristic.

$$Q = \frac{\omega_0}{\text{BW}} \tag{8.20}$$

Hence the Q sensitivity S_x^Q is directly related to how sensitive the bandwidth is to an element value change.

The Q and ω_0 sensitivities are frequently extended to any quadratic polynomial even if it is that of a denominator of a lowpass, highpass, or bandreject filter function. They can also be applied to a quadratic polynomial even if it is a numerator.

In other situations, a second-order network function may be written as

$$H(s) = \frac{Ks}{(s - s_p)(s - s_p^*)} \tag{8.21}$$

We may be interested in the sensitivity of the pole position - s_p or s_p^* - with respect to change in a certain element value. The quantity $S_x^{s_p}$ is known as the pole sensitivity. This notion can also be extended to a quadratic numerator. In that case, the sensitivity would be the zero sensitivity.

8.4   Sensitivity calculation

Given a circuit, the sensitivity of one of its performance measures, y, with respect to change in a certain element value, x, can always be calculated directly by first expressing y in terms of x. Then the definition given in (8.2) can be used to calculate the sensitivity directly. The step of expressing y in terms of x can be quite laborious. Frequently, it is more expedient to obtain the sensitivity in several substeps using some of the properties given in Section 8.2.

For example, to obtain the transfer function of the active filter of Fig. 8.4, we can write two node equations - one for the node whose voltage is E_3

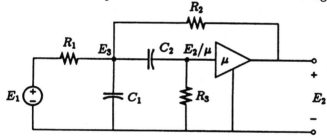

Figure 8.4: An active bandpass filter circuit.

and one for the input node of the ideal amplifier with the voltage E_2/μ. Using MATLAB, we will have the following steps.

```
» f1 = '(E3-E1)/R1 + s*C1*E3 + s*C2*(E3-E2/u)
+ (E3-E2)/R2';
» f2 = 's*C2*(E2/u-E3) + E2/(u*R3)';
» pretty(solve(f1,f2,'E2,E3'))
E2 = s C2 R3 u R2 E1/(- s C2 R3 R1 u + R2 s C2 R3
+ R2 + s C1 R1 R2 C2 R3 + s C1 R1 R2 + s C2 R1 R2
+ R1 s C2 R3 + R1),
E3 = (s C2 R3 + 1) R2 E1/(- s C2 R3 R1 u + R2 s C2 R3
+ R2 + s C1 R1 R2 C2 R3 + s C1 R1 R2 + s C2 R1 R2
+ R1 s C2 R3 + R1)
```

Here, we solved for both E_2 and E_3 in order to eliminate the latter. The above result gives

$$H(s) = \frac{E_2}{E_1} = \frac{Ks}{s^2 + b_1 s + b_0}$$

$$= \frac{\dfrac{\mu}{C_1 R_1} s}{s^2 + \left(\dfrac{1}{C_2 R_3} + \dfrac{1}{C_1 R_3} + \dfrac{1}{C_1 R_1} + \dfrac{1 - \mu}{C_1 R_2}\right) s + \dfrac{R_1 + R_2}{C_1 C_2 R_1 R_2 R_3}} \quad (8.22)$$

If we wished to obtain the pole-Q sensitivity, we would have to deal with

$$Q = \frac{\sqrt{\dfrac{R_1 + R_2}{C_1 C_2 R_1 R_2 R_3}}}{\dfrac{1}{C_2 R_3} + \dfrac{1}{C_1 R_3} + \dfrac{1}{C_1 R_1} + \dfrac{1 - \mu}{C_1 R_2}} \quad (8.23)$$

Instead, since $Q = \sqrt{b_0}/b_1$, using (8.10) and (8.12) we can use the relationship

$$S_x^Q = S_x^{\sqrt{b_0}} - S_x^{b_1} = \frac{1}{2} S_x^{b_0} - S_x^{b_1} \quad (8.24)$$

This alternative approach is also preferable as $S_x^{b_1}$ and $S_x^{b_0}$ can also used in other sensitivity calculations as well.

Specifically, if we make $C_1 = C_2 = 1$ F, $\mu = 2$, $R_1 = 4 \ \Omega$, $R_2 = 0.7255 \ \Omega$, and $R_3 = 1.6283 \ \Omega$, the circuit of Fig. 8.4 would realize

$$H(s) = \frac{0.5s}{s^2 + 0.1s + 1} \quad (8.25)$$

Suppose we wish to calculate several sensitivities with respect to C_1. We first find

$$S_{C_1}^{b_1} = \frac{\dfrac{1}{C_1 R_3} + \dfrac{1}{C_1 R_1} + \dfrac{1 - \mu}{C_1 R_2}}{\dfrac{1}{C_2 R_3} + \dfrac{1}{C_1 R_3} + \dfrac{1}{C_1 R_1} + \dfrac{1 - \mu}{C_1 R_2}} \cdot S_{C_1}^{\frac{1}{C_1 R_3} + \frac{1}{C_1 R_1} + \frac{1 - \mu}{C_1 R_2}}$$

$$= -\frac{\dfrac{1}{C_1 R_3} + \dfrac{1}{C_1 R_1} + \dfrac{1-\mu}{C_1 R_2}}{\dfrac{1}{C_2 R_3} + \dfrac{1}{C_1 R_3} + \dfrac{1}{C_1 R_1} + \dfrac{1-\mu}{C_1 R_2}} = 5.1466$$

$$S_{C_1}^{b_0} = -1$$

$$S_{C_1}^{\omega_0} = S_{C_1}^{\sqrt{b_0}} = -\frac{1}{2}$$

$$S_{C_1}^{Q} = S_{C_1}^{\omega_0/b_1} = -\frac{1}{2} - S_{C_1}^{b_1} = -5.6466$$

$$S_{C_1}^{\sqrt{b_1^2 - 4b_0}} = \frac{1}{2} S_{C_1}^{b_1^2 - 4b_0} = \frac{1}{2} \cdot \frac{2 b_1^2 S_{C_1}^{b_1} - 4 b_0 S_{C_1}^{b_0}}{b_1^2 - 4b_0} = -0.51415$$

Of course, any sensitivity can also be obtained directly if the expression of the quantity is available. Usually, this effort will require the aid of some computer software. For example, suppose we wish to obtain $S_{C_1}^{Q}$. We can apply the following MATLAB steps on the expression of Q given in (8.23).

```
» Q = '(sqrt((R1+R2)/(C1*C2*R1*R2*R3))/
        (1/(C2*R3) +1/(C1*R3)+1/(C1*R1)+(1-u)/(C1*R2)))';
% Substitute the element values
» Q1 = SUBS(Q,1,'C1');
» Q2 = SUBS(Q1,1,'C2');
» Q3 = SUBS(Q2,2,'u');
» Q4 = SUBS(Q3,4,'R1');
» Q5 = SUBS(Q4,0.7255,'R2');
» Q6 = SUBS(Q5,1.6283,'R3');
» Qval = eval(Q6)
Qval = 10.0087
% Differentiate Q with respect to C1
» dQ = DIFF(Q,'C1');
% Substitute the element values
» dQ1 = SUBS(dQ0,1,'C1');
» dQ2 = SUBS(dQ1,1,'C2');
» dQ3 = SUBS(dQ2,2,'u');
» dQ4 = SUBS(dQ3,4,'R1');
» dQ5 = SUBS(dQ4,0.7255,'R2');
» dQ6 = SUBS(dQ5,1.6283,'R3');
» dQval = eval(dQ6) dQval = -56.5149
```

```
> C1 = 1;
% Use Equation (8.2) to evaluate the sensitivity
> SEN = C1/Qval*dQval
SEN = -5.6466
```

which agrees with the calculation done previously.

For the pole sensitivity, since $s_p = \frac{1}{2}(-b_1 + \sqrt{b_1^2 - 4b_0})$, we can say that

$$S_{C_1}^{s_p} = \frac{-b_1 S_{C_1}^{-b_1} + \sqrt{b_1^2 - 4b_0} S_{C_1}^{\sqrt{b_1^2 - 4b_0}}}{-b_1 + \sqrt{b_1^2 - 4b_0}} = -0.5 + j0.28268$$

For the sensitivity of $|H(j)|$, since

$$|H(j)| = \frac{\dfrac{\mu}{C_1 R_1}}{\sqrt{(b_0 - 1)^2 + b_1^2}}$$

we have

$$S_{C_1}^{|H(j)|} = -1 - \frac{1}{2} S_{C_1}^{[(b_0-1)^2 + b_1^2]} = -1 - \frac{1}{2} \cdot \frac{(b_0 - 1)^2 S_{C_1}^{(b_0-1)^2} + b_1^2 S_{C_1}^{b_1^2}}{(b_0 - 1)^2 + b_1^2}$$

$$= -1 - \frac{(b_0 - 1)^2 S_{C_1}^{(b_0-1)} + b_1^2 S_{C_1}^{b_1}}{(b_0 - 1)^2 + b_1^2} = -1 - S_{C_1}^{b_1} = -6.1466$$

8.5 Unnormalized sensitivity

The differential sensitivity is a normalized quantity in that both Δy and Δx are compared with y and x respectively. The normalized sensitivity associated with certain quantities of a network function is not as meaningful as the unnormalized sensitivity. One such quantity is the position of the poles or zeros (collectively called roots). We are usually more interested in how much a root actually migrates as one or more element values are varied. For example, in the case of the pole sensitivity calculation done in the previous section,

$$S_{C_1}^{s_p} \approx \frac{\frac{\Delta s_p}{s_p}}{\frac{\Delta C_1}{C_1}} \tag{8.26}$$

is not very informative since we should be more concerned about how much s_p changes rather than its per-unit change.

The unnormalized sensitivity is defined as

$$US_x^y = x \cdot \frac{dy}{dx} \tag{8.27}$$

Obviously,

$$US_x^y = yS_x^y \quad \text{and} \quad \Delta y \approx \frac{\Delta x}{x} US_x^y \tag{8.28}$$

Therefore computations of the normalized sensitivity and the unnormalized sensitivity are quite similar.

In the example of the previous section, the unnormalized pole sensitivity is

$$US_{C_1}^{s_p} = s_p S_{C_1}^{s_p} = -0.25733 - j0.51351$$

Hence, we may estimate that if C_1 is increased by 1%, the pole will move from $-0.050214 + j0.99826$ to

$$-0.050214 + j0.99826 + 0.01 \times (-0.25733 - j0.51351)$$

$$= -0.052788 + j0.99312$$

Actual computation with $C_1 = 1.01$ F will give the new pole position to be $-0.052503 + j0.99367$. The estimated pole position compares well with the actual location.

8.6 Multiparameter and statistical sensitivities

In the design of a filter, particularly those using active devices, we sometimes focus our attention on the sensitivity of a certain performance

measure with respect to a certain network element value. Effort may be
made to minimize this particular sensitivity, primarily because we know
a priori that this particular sensitivity is the most critical one. Or else,
since we have a wide choice of combinations of element values to give us
the same nominal network function, we can take advantage of this fact
and minimize one of the sensitivities.

In practice, it can be assumed that the performance of a filter is affected
by all element changes. In general terms, we can assume that y is a
function of several x's and we expresses $y = f(x_1, x_2, \ldots, x_n)$. Then
there will be n sensitivities for each y as defined in (8.3). If we only keep
the first-order changes, we can write

$$\frac{\Delta y}{y} \approx \sum_{i=1}^{n} S_{x_i}^y \cdot \frac{\Delta x_i}{x_i} \tag{8.29}$$

In some situations, the elements may be divided into several groups
according to their types. Then each type of elements may be assumed
to vary by the same extent. For example, in an active filter, there may
be a number of resistances, $R_1, R_2, \ldots, R_m$, a number of capacitances,
$C_1, C_2, \ldots, C_k$, and a number of amplifiers with gains, $\mu_1, \mu_2, \ldots, \mu_j$.
Then the per-unit variation of y may be written as

$$\frac{\Delta y}{y} \approx \left(\sum_{i=1}^{m} S_{R_i}^y \right) \frac{\Delta R}{R} + \left(\sum_{i=1}^{k} S_{C_i}^y \right) \frac{\Delta C}{C} + \left(\sum_{i=1}^{j} S_{\mu_i}^y \right) \frac{\Delta \mu}{\mu} \tag{8.30}$$

It is conceivable that each term of (8.30) may be relatively large in absolute value while their combination may be quite small as the change of
one group can be offset by another. Even within one group, it is possible that some of the sensitivities have opposite signs and the composite
effects neutralize one another.

When a large number (say, hundreds of thousands) of filters are to be
fabricated, a consideration related to sensitivity needs to be considered.
In such a situation, we assume each element to vary according to some
statistical distribution pattern such as its probability density function.
Based on this assumption, we determine the statistical distribution of
the performance of the filters. Those that fall within certain bounds
are considered acceptable. The number of acceptable units determines
the yield of the fabrication process. The study of how the required
tolerances of the various elements should be assigned in order to predict
statistically the percent yield of the process is known as *tolerancing*.
Computer simulation is often used to determine the tolerances of the

various component values in order to improve the percent yield so the fabrication process can be economically justified.

8.7   A case for low sensitivity of passive filters

Engineers have long believed and experiences have often affirmed that active filters are usually more sensitive then passive ones. Unfortunately this impression cannot be proved theoretically. In fact, counter-examples can be found to dispute this conjecture in general terms. However, in practice, this belief is generally accepted. One of the reasons for the low sensitivity of passive filters is that, as a rule, all elements in a passive filter circuit have comparable effects on the network performance. Hence the burden of accomplishing the filtering task is shared by all elements more uniformly than in an active filter. The other reason is the absence of amplification in the circuit. Hence the effect of any element value change is not somehow unduly exacerbated by some active devices.

However, one situation can be used to show that the sensitivity of passive filters can be expected to be extremely low. This is the doubly-terminated LC twoport. If we design the twoport such that at some points in the pass band, $|t(j\omega_j)| = 1$, and since $|t(j\omega)| \leq 1$, any change in the passive element values, in either direction, can only decrease $|t(j\omega_j)|$. This is illustrated in Fig. 8.5 in which x_0 represents the nominal value of a typical element. Hence

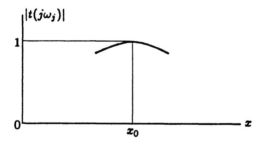

Figure 8.5: Variation of $|t(j\omega_j)|$ as x is varied in the neighborhood of x_0.

$$\frac{d(|t(j\omega_j)|)}{dx}\bigg|_{x=x_0} = 0 \quad \Longrightarrow \quad S_{x_0}^{|t(j\omega_j)|} = 0 \qquad (8.31)$$

If the passband ripple is small, say, in the order of 0.1 dB, we can expect the sensitivity throughout the pass band to be low because of the unity upper bound of $|t(j\omega)|$.

Similar expectations cannot be held for doubly-terminated LC filters in the stop band, or for singly-terminated LC filters.

8.8 Summary

In this chapter we have introduced the subject of sensitivity. We have chosen to mention this subject at this point, because it is an important collateral consideration for active filters, which we will begin to explore in the coming chapters. We have given the classical definition of sensitivity, developed its mathematical properties, and demonstrated typical methods of computing it.

We will have occasion to show that sometimes certain active filters can be optimized with respect to certain important sensitivities. We can use the comparison of sensitivities of various circuits to determine which circuits are superior in certain respects.

Problems

8.1 Derive the properties given in (8.11) through (8.17) in Section 8.2.

8.2 Show that

$$\frac{d}{dx}S_x^y = \frac{1}{x}S_x^y\left(1 + S_x^{dy/dx} - S_x^y\right)$$

8.3 For the following circuit, compute $S_L^{|t(j)|^2}$ and $S_C^{|t(j)|^2}$. Compare your answers with the approximate expression given by (8.4) by letting $\Delta x/x = 0.01$.

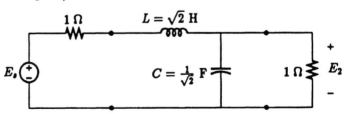

8.4 For the following circuit, let $H(s) = I/E$. (a) Obtain the sensitivity $S_C^{|H(j\omega)|^2}$ at $\omega = 1/\sqrt{LC}$. (b) Let $L = 1$ H, $C = 1$ F, and $R = 0.01\ \Omega$. Calculate the actual value of

$$\frac{C}{|H(j\omega)|^2} \times \frac{\Delta|H(j\omega)|^2}{\Delta C}$$

for a 1% increase in the value of C ($\Delta C/C = 0.01$).

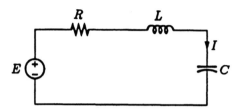

8.5 Suppose

$$H(s) = \frac{Ks}{s^2 + \left(\dfrac{\omega_0}{Q}\right)s + \omega_0^2} = \frac{Ks}{s^2 + \dfrac{1}{R_1 C_1}s + \dfrac{1}{R_1 R_2 C_1 C_2}}$$

For $C_1 = C_2 = 1$ F, $R_1 = 4$ Ω, and $R_2 = 1$ Ω, calculate the sensitivities of Q and ω_0 with respect to each of the four passive elements.

8.6 Given

$$H(s) = \frac{Ks}{s^2 + \left(\dfrac{\omega_0}{Q}\right)s + \omega_0^2} = \frac{Ks}{s^2 + \left(\dfrac{1}{R_1 C_1} + \dfrac{1-\mu}{R_2 C_2}\right)s + \dfrac{1}{R_1 R_2 C_1 C_2}}$$

derive the sensitivities of Q and ω_0 with respect to μ.

8.7 For a given lowpass second-order network function

$$H(s) = \frac{\omega_0^2}{s^2 + \left(\dfrac{\omega_0}{Q}\right)s + \omega_0^2}$$

Let $G(\omega) = 20 \log |H(j\omega)|$. Derive the expression for $S_Q^{G(\omega)}$ and $S_{\omega_0}^{G(\omega)}$. Then evaluate these sensitivities at $\omega = \omega_0$.

8.8 Calculate Q, ω_0, and $|H(j1)|$ sensitivities for the example used in Section 8.4 (Fig. 8.4 and (8.25)) when x is R_3 instead of C_1.

8.9 Obtain the voltage ratio $H(s) = E_2/E_1$ of the following circuit in which $R_1 = 1.6$ Ω, $R_2 = 3.2$ Ω, $R_3 = 5.5$ Ω, $C_1 = 0.05$ F, and

$C_2 = 1$ F. Then calculate the sensitivities of Q, ω_0, and $|H(j1)|$ with respect to each passive element.

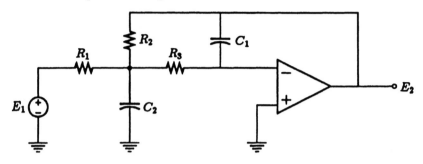

8.10 For the circuit of the previous problem, calculate the unnormalized pole sensitivities with respect to each passive element.

Chapter 9

Basics of active filters

Chapters 5, 6, and 7 deal with the theory and basic methods of realizing filters that use passive elements - chiefly inductors and capacitors. These filters were the mainstay for filtering applications from the 1920's through the 1940's. Since the 1940's, another type of filters - the active filters - have emerged and the technology has improved to the point that they are now in very common use and have attained a great deal of success. Originally motivated mainly by the low-frequency applications of filters in which inductors are too costly or too bulky and heavy, engineers began to search for alternative approaches using active devices. In the early stages of the development of active filters, some successes were achieved even when vacuum tubes were used. However, those filters were only feasible for very low frequency applications. When low-cost, light-weight, low-voltage solid-state devices became available, active filters became much more attractive and are applicable over much wider frequency ranges. As integrated-circuit devices - chiefly the op amp - became more and more economical and their applicable frequency range became wider and wider, so active filters became quite competitive with passive ones. Nowadays, both types of filters have their appropriate places in filtering applications.

9.1 Comparison of passive and active filters

Except for applications at extremely low or extremely high frequencies, active and passive filters are both useful in many applications. However, there are some fundamental differences between these two types of filters.

Obviously, every active filter requires a power supply, while a passive one does not. This may or may not be a determining factor in deciding which type of filter to use. In many applications, a power supply is already available for other purposes.

The terminating resistances, whether singly terminated or doubly terminated, are an integral part of the passive filter circuit. This requires the synthesis procedure of a passive filter circuit to be very precise and restrictive. As a result, the development of the procedure is usually much more difficult. Another consequence of this condition is that the number of available circuits to realize a given network function is very limited - sometimes even unique. In contrast, active filters are usually designed without regard to the load or source impedance. Since buffer amplifiers are available, the source, the filter, and the terminating impedance can be isolated from one another. Also, in many active filter circuits, the amplifier output terminal is also the output of the filter. Hence the value of the terminating impedance does not affect the performance of the filter.

Partly because of this feature and partly because of the wide availability of different active devices and circuits, the number of available active circuits to realize a given network function can be very large. In the early development of active filters, literally hundreds of different circuits were proposed, some of them slight variations of existing ones while others were obtained by many novel approaches.

It is generally accepted that passive filters are less sensitive to element value variations than active ones. This assertion is partly supported by the phenomenon described in Section 8.7.

The use of active circuits in filtering applications also brings with it several inherent problems that are not serious in passive filters. The active circuits are typically more noisy, they have limited dynamic ranges, and they are prone to instability due to parasitic positive feedbacks.

Because inductors of any significant value cannot be fabricated by integrated circuits, passive filters are generally produced in discrete or hybrid form. Modern active filters employ largely resistors, capacitors, and op amps. They can be fabricated in thick-film, thin-film, or integrated-circuit form. New schemes that take advantage of new material properties and new solid-state circuit devices are still being developed.

The study, synthesis, and implementation of active filters enjoys another feature that passive ones do not. That is the possibility of interconnecting simple standard building blocks to form complicated filters. This makes it possible to mass-produce active circuit modules, sometimes with a few externally supplied components, to perform many different filtering tasks. For example, if we wish to realize a sixth-order Butterworth lowpass filter with

$$|H(j\omega)|^2 = \frac{1}{1 + \omega^{12}} \tag{9.1}$$

with passive networks, we are forced to realize

$$H(s) = 1/(s^6 + 3.863703s^5 + 7.464102s^4$$

$$+9.141620s^3 + 7.464102s^2 + 3.863703s + 1) \tag{9.2}$$

in one step. In active realization, we can realize three second-order sections such that (see Tables A.1 and A.2)

$$H_1(s) = \frac{1}{s^2 + 0.517638s + 1}$$

$$H_2(s) = \frac{1}{s^2 + 1.414214s + 1} \tag{9.3}$$

$$H_3(s) = \frac{1}{s^2 + 1.931852s + 1}$$

and connect these second-order sections in cascade as shown in Fig. 9.1. In active filters, second-order sections are referred to as *biquads*.

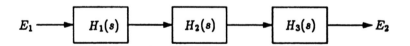

Figure 9.1: Cascade of three biquads.

On other occasions, lower-order sections may be interconnected by feedback or feed-forward arrangements. An example is the feedback arrangement shown in Fig. 9.2. To realize the $H(s)$ in (9.2), the three biquads can be made identical.

In this text, we'll limit our study to active filters that employ resistors, capacitors, and op amps as their circuit elements. We'll first describe the op amp and some simple circuits using these elements in this chapter. In later chapters, we'll describe some building blocks, how they can be interconnected to form complex filters, and other methods of realizing active filters.

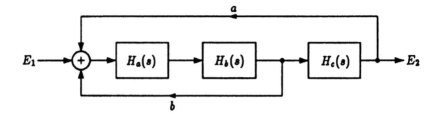

Figure 9.2: Three cascaded biquads with feedback.

9.2 The operational amplifier (op amp)

For our purposes, we shall regard an op amp as a differential-input, grounded-output amplifier as shown in Fig. 9.3. In an ideal op amp, we further assume:

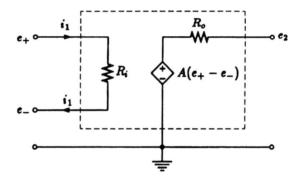

Figure 9.3: Equivalent circuit of an op amp.

(1) The input resistance, R_i, approaches infinity. Thus, $i_1 = 0$.

(2) The output resistance, R_o, approaches zero.

(3) The amplifier gain A, approaches infinity.

The third assumption requires $e_+ = e_-$ as e_2 must be finite. This condition between the two terminals is said to be a *virtual short*. If one of the terminals is physically grounded, then the other terminal is a *virtual ground*.

An idealized op amp is represented by the symbol shown in Fig. 9.4. The terminal connected to the '+' sign has the voltage e_+ and is known as the *noninverting terminal*. The terminal connected to the '−' sign has the voltage e_- and is known as the *inverting terminal*. The ground

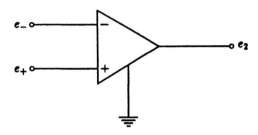

Figure 9.4: Symbol for an ideal op amp.

is frequently omitted in a circuit diagram. But it is always implied.

Two practical limitations of the op amp need to be mentioned here. The assumptions made on the ideal op amp imply a high-gain, infinite-bandwidth amplifier. Such an amplifier, with its parasitics, is inherently unstable. Therefore, all op amps have to have some means to limit their bandwidths and high-frequency gains. Usually a compensating circuit is used for this purpose. As a first-order approximation, the op amp gain is a function of s and

$$A(s) = \frac{A_0}{1 + \dfrac{s}{\omega_a}} \tag{9.4}$$

This is the primary reason why active filters either are not feasible or will not be cost-effective at very high frequencies.

The other limitation is that its maximum signal swing is obviously limited. Certainly, the peak value of the signal in any part of the interior of an op amp cannot exceed its power supply voltage. To ensure little or no distortion, the signal strength should be kept well below the dc power-supply level. At the other extreme, since an op amp is a relatively noisy device, the signal level should not be too weak anywhere in an active filter, so that the noise generated by the active devices will not corrupt the signal too much to interfere with its transmission.

Since the theme of this text is the use of the op amp as a filter component, we will not deal with the many practical facets of op amps. The reader is encouraged to take courses dealing with op amps and their properties. We shall regard every op amp we use as ideal.

An op amp is never used in the open-loop mode. Some form of feedback is used to make the closed-loop gain finite or a function of s. In closing the loop it is always critical which of the two input terminals is connected to the feedback circuit.

9.3   Some simple op amp circuits

Figures 9.5 through 9.12 present some simple op amp circuits that will be useful in active filter circuits.

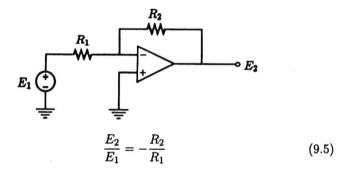

$$\frac{E_2}{E_1} = -\frac{R_2}{R_1} \tag{9.5}$$

Figure 9.5: Inverting voltage amplifier.

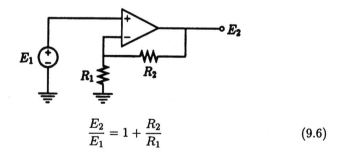

$$\frac{E_2}{E_1} = 1 + \frac{R_2}{R_1} \tag{9.6}$$

Figure 9.6: Noninverting voltage amplifier.

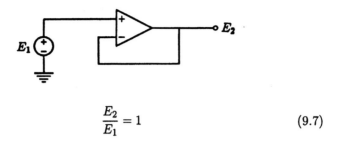

$$\frac{E_2}{E_1} = 1 \qquad (9.7)$$

Figure 9.7: Voltage follower (buffer amplifier).

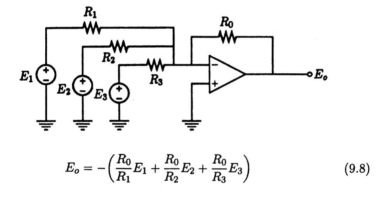

$$E_o = -\left(\frac{R_0}{R_1}E_1 + \frac{R_0}{R_2}E_2 + \frac{R_0}{R_3}E_3\right) \qquad (9.8)$$

Figure 9.8: Inverting weighted summer.

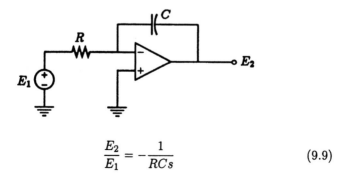

$$\frac{E_2}{E_1} = -\frac{1}{RCs} \qquad (9.9)$$

Figure 9.9: Inverting lossless integrator.

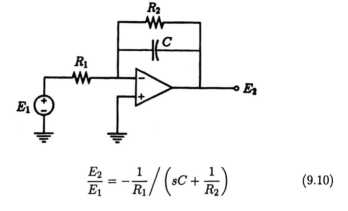

$$\frac{E_2}{E_1} = -\frac{1}{R_1} \bigg/ \left(sC + \frac{1}{R_2}\right) \qquad (9.10)$$

Figure 9.10: Inverting lossy integrator.

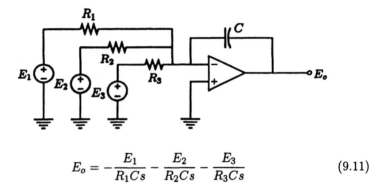

$$E_o = -\frac{E_1}{R_1 C s} - \frac{E_2}{R_2 C s} - \frac{E_3}{R_3 C s} \qquad (9.11)$$

Figure 9.11: Inverting weighted summing integrator.

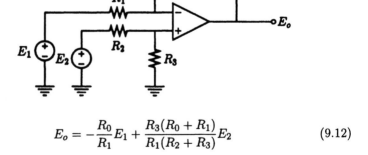

$$E_o = -\frac{R_0}{R_1} E_1 + \frac{R_3(R_0 + R_1)}{R_1(R_2 + R_3)} E_2 \qquad (9.12)$$

Figure 9.12: Differential weighted summer (subtractor).

9.4 First-order sections

In realizing an odd-ordered voltage transfer function using the cascade connection, it is necessary that one of the sections be odd-ordered. The simplest odd-ordered section is the first-order section. The general expression for the inverting first-order section is

$$\frac{E_2}{E_1} = -\frac{a_1 s + a_0}{s + b_0} \tag{9.13}$$

Now consider the arrangement of Fig. 9.13(a), which is the generalized version of the inverting amplifier of Fig. 9.5. We have

$$\frac{E_2}{E_1} = -\frac{Z_2}{Z_1} = -\frac{Y_1}{Y_2} \tag{9.14}$$

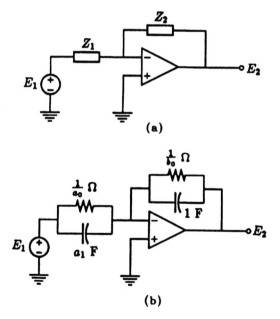

(a)

(b)

Figure 9.13: General inverting first-order section.

Equating the numerators and the denominators of (9.13) and (9.14), we have

$$Y_1 = a_1 s + a_0 \qquad \text{and} \qquad Y_2 = s + b_0$$

Thus the circuit of Fig. 9.13(b) realizes the general inverting first-order section of (9.13). Note, if $a_1 = 0$, the circuit becomes an inverting lossy integrator as in Fig. 9.10. If $a_1 = 0$ and $b_0 = 0$, the circuit becomes an inverting lossless integrator as in Fig. 9.9.

If a noninverting first-order section is desired, the transfer function is

$$\frac{E_2}{E_1} = \frac{a_1 s + a_0}{s + b_0} \tag{9.15}$$

We can use the arrangement of Fig. 9.14(a), which is the generalized version of the noninverting amplifier of Fig. 9.6. We have

$$\frac{E_2}{E_1} = 1 + \frac{Z_2}{Z_1} = 1 + \frac{Y_1}{Y_2} \tag{9.16}$$

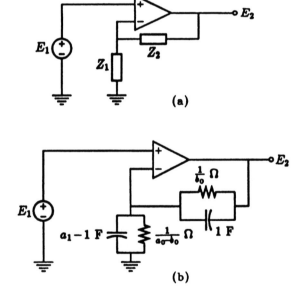

(a)

(b)

Figure 9.14: General noninverting first-order section.

Equations (9.15) and (9.16) require that

$$\frac{Y_1}{Y_2} = \frac{a_1 s + a_0}{s + b_0} - 1 = \frac{(a_1 - 1)s + (a_0 - b_0)}{s + b_0} \tag{9.17}$$

If $a_1 > 1$ and $a_0 > b_0$, the circuit of Fig. 9.14(b) realizes (9.15). These two inequalities can always be satisfied by multiplying the numerator of right-hand side of (9.15) by a sufficiently large constant. In fact, if we make one of the inequalities into an equality, one element will be saved. This multiplication amounts to a change in the overall gain of the section.

There are numerous other circuits that will realize the first-order transfer function. Some of these circuits will be found in the problems at the end of the chapter.

9.5 RC single-op amp circuit relationships

Many biquad circuits use a combination of a grounded RC threeport and an op amp; the latter may be used either as an ideal op amp or to form a voltage amplifier. We shall treat the general mathematical relationships of two of these arrangements.

9.5.1 Finite-gain single-op amp configuration

This configuration is shown in Fig. 9.15. The voltage amplifier can in turn be realized by using the circuit of Fig. 9.6. The RC grounded threeport is assumed to have the y parameters which relate its currents and voltages by the general equations

$$I_1 = y_{11}E_1 + y_{12}E_2 + y_{13}E_3$$

$$I_2 = y_{21}E_1 + y_{22}E_2 + y_{23}E_3 \tag{9.18}$$

$$I_3 = y_{31}E_1 + y_{32}E_2 + y_{33}E_3$$

Since the amplifier is assumed to be ideal, we have $I_3 = 0$. Also, the amplifier requires that $E_3 = E_2/\mu$. Substituting these into the third equation of (9.18), we get

$$0 = y_{31}E_1 + y_{32}E_2 + \frac{y_{33}}{\mu}E_2$$

We rearrange to obtain

$$\frac{E_2}{E_1} = -\frac{y_{31}}{y_{32} + \dfrac{y_{33}}{\mu}} \tag{9.19}$$

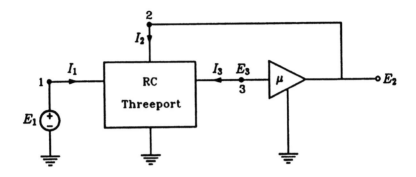

Figure 9.15: RC threeport with one voltage amplifier.

In most active biquad circuits, the RC threeports are not very complicated. Their driving-point admittances typically have only two poles. Their transfer admittances may have fewer than two poles. Equation (9.19) enables us to infer that the transmission zeros of the biquad are the zeros of y_{31} and the private poles of y_{33}. Thus, to determine the transmission zeros of the arrangement of Fig. 9.15, we can first determine the transmission zeros of the RC threeport between port 1 and port 3. Since

$$y_{31} = \frac{I_3}{E_1}\bigg|_{E_2=E_3=0} \qquad (9.20)$$

to determine the transmission zeros between port 1 and port 3, we can investigate the RC threeport arrangement shown in Fig. 9.16. If for any value of s, $I_3 \equiv 0$, then that value of s must be a transmission zero of the biquad. If the number of zeros of y_{31} is less than two, then we need to investigate further the poles of y_{33}. The poles in y_{33} that are not also poles of y_{31} will also be the transmission zeros of the biquad.

For example, the transmission zeros of the biquad of the filter shown in Fig. 9.17(a) can be determined by examining the zeros of y_{31} and the poles of y_{33} of the passive threeport in Fig. 9.17(a). To determine the zeros of y_{31}, we examine the arrangement of Fig. 9.17(b).

As $s \to 0$, capacitor C_1 approaches an open circuit. Therefore, y_{31} has a zero at $s = 0$. This happens to be the only zero. Although C_2 approaches a short circuit at $s \to \infty$, it does not produce a zero since it is in parallel with C_1. Hence it is necessary to determine the poles of y_{33}, as depicted in Fig. 9.17(c). From this arrangement, it is easily seen that y_{33} has a private pole at $s \to \infty$ since C_1 and C_2 both approach short circuits at

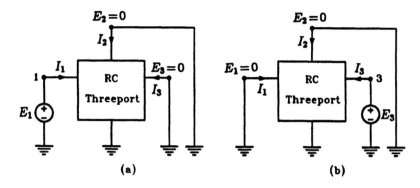

Figure 9.16: Arrangements used to determine the zeros of y_{31}.

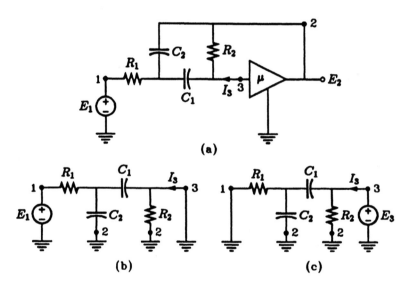

Figure 9.17: An example of the determination of transmission zeros.

high frequencies. Hence, the filter must have a transmission zero at the origin and another at infinity. The transfer function of the biquad must have the form

$$\frac{E_2}{E_1} = \frac{Ks}{s^2 + b_1 s + b_0} \qquad (9.21)$$

9.5.2   Infinite-gain single-op amp configuration

The transfer function of the arrangement of Fig. 9.18 can be obtained
by letting $\mu \to \infty$ in (9.19) or

$$\frac{E_2}{E_1} = -\frac{y_{31}}{y_{32}} \tag{9.22}$$

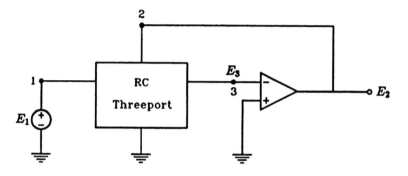

Figure 9.18: RC threeport with an op amp.

Equation (9.22) also shows that the transmission zeros of the biquad are
the same as the zeros of y_{31} of the RC threeport. Also, it can be seen that
the poles of the transfer function are the zeros of y_{32} - the short-circuit
transfer admittance between port 2 and port 3 of the threeport. This can
easily be explained by the fact that when y_{32} is zero, there is no feedback
from the output to the inverting terminal of the op amp. Therefore, at
frequencies where there is no negative feedback, E_2 becomes unbounded.

9.6   Gain adjustments

Situations do arise when the gain of an active filter section is either too
high or too low. It can be argued that since amplifiers are available, this
is not a problem with active filters. However, there are considerations
other than the overall gain. If the signal is too high at some points of a
filter, either the op amp will clip the signal or the device will be latched
to one of its power supply voltages. On the other hand, if the signal is too
low, the noise in the circuit may cause the filter to perform erratically.
Hence, it may be necessary to adjust the gain of an individual section
to the appropriate level. This can be accomplished by the methods
described in this section.

9.6.1 Gain reduction

To reduce the gain of a filter section, we can use the scheme shown in Fig. 9.19. We replace every impedance that is connected to the input terminal by a voltage divider while keeping the Thévenin equivalent impedance unchanged. Network N represents the remainder of the filter

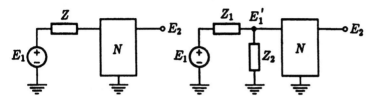

Figure 9.19: Method of gain reduction.

section. Suppose we wish to reduce the gain to α times its original value, where $\alpha < 1$. We make

$$\frac{E_1'}{E_1} = \alpha = \frac{Z_2}{Z_1 + Z_2} \quad \text{and} \quad \frac{Z_1 Z_2}{Z_1 + Z_2} = Z \tag{9.23}$$

Solving for Z_1 and Z_2, we get

$$Z_1 = \frac{Z}{\alpha} \quad \text{and} \quad Z_2 = \frac{1}{1 - \alpha} Z \tag{9.24}$$

As an example, the filter of Fig. 9.20 realizes the voltage transfer function

$$\frac{E_2}{E_1} = \frac{4s^2 + 1}{4s^2 + 4s + 1} \tag{9.25}$$

Suppose we wish the reduce the gain by 25% or $\alpha = 0.75$. We apply (9.24) to both C_1 and R_1. The circuit of Fig. 9.21 results.

9.6.2 Gain enhancement

If it is desired to increase the gain of a filter and if a voltage amplifier is present at the output terminal, a simple scheme is to increase the amplifier gain and decrease the feedback by the same amount. Suppose the original filter has the configuration of Fig. 9.15. We can increase the amplifier gain to $K\mu$ and, at the same time, decrease the feedback by K as shown in Fig. 9.22.

For the RC threeport, we have

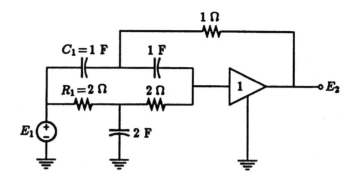

Figure 9.20: Filter that realizes the transfer function of (9.25).

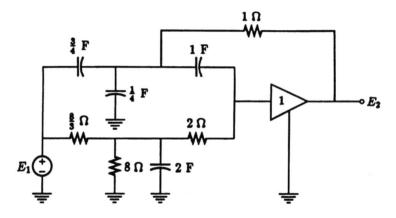

Figure 9.21: Filter of Fig. 9.20 modified to reduce its gain.

$$y_{31} E_1 + y_{32} \frac{E_2}{K} + y_{33} \frac{E_2}{K\mu} = 0$$

which gives

$$\frac{E_2}{E_1} = -\frac{y_{31}}{\dfrac{y_{32}}{K} + \dfrac{y_{33}}{K\mu}} = -\frac{K y_{31}}{y_{32} + \dfrac{y_{33}}{\mu}} \qquad (9.26)$$

which shows an increase in gain by a factor K over what is given in (9.19).

Since an amplifier with a gain of $1/K$ is really an attenuator (if $K > 1$) it can be replaced by a passive voltage divider exactly as was done to the input voltage in the gain-reduction technique shown in Fig. 9.19 with (9.24).

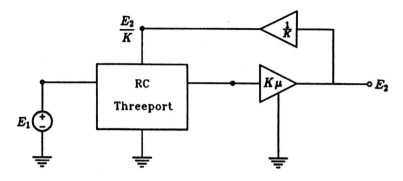

Figure 9.22: Arrangement for gain enhancement.

As an example, suppose we wish to increase the gain of the filter in Fig. 9.20 by a factor of 2. We can use either of the two circuits shown in Fig. 9.23.

If the output amplifier is an ideal infinite-gain op amp, then to enhance the gain will only require a reduction in feedback. This is because y_{33} does not appear in the denominator of (9.22). Hence a reduction in y_{32} causes a proportionate increase in the filter gain. This feedback reduction can be effected either by using an amplifier with a gain less than unity or by using the method shown in Fig. 9.19 with (9.24).

9.7   RC-CR transformation

The frequency transformations we discussed in Chapter 4 are useful in obtaining highpass, bandpass, and bandreject network functions from that of a lowpass prototype. If the lowpass filter is an RC-op amp filter, the direct application of these transformations to circuit elements as was summarized in Table 4.1 will require the use of inductors. Therefore these transformations cannot be used to obtain highpass, bandpass and bandreject active filters from a lowpass filter prototype.

There is one transformation that is useful in transforming certain active filters [Mi]. This transformation is known as the RC-CR transformation. This transformation is applicable to a network N that contains resistors, capacitors, and dimensionless controlled sources (voltage-controlled voltage sources or current-controlled current sources.) It has been shown that if every conductance of G_i mhos in N is replaced with a capacitance of G_i farads and every capacitance in N of C_j farads is replaced with a conductance of C_j mhos, to form a new network N', the following

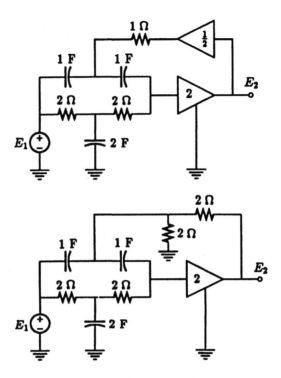

Figure 9.23: Examples of gain enhancement.

conclusions can be drawn on the relationship between N and N'. (Every unprimed function is that of N and every primed function is that of N'.)

(1) The corresponding network functions with the dimension of the impedance (e.g. E_2/I_1 and E_1/I_1) must satisfy

$$Z'(s) = \frac{1}{s} Z \left(\frac{1}{s} \right) \tag{9.27}$$

(2) The corresponding network functions with the dimension of the admittance (e.g. I_2/E_1 and I_1/E_1) must satisfy

$$Y'(s) = sY \left(\frac{1}{s} \right) \tag{9.28}$$

(3) The corresponding network functions that are dimensionless (e.g. E_2/E_1 and I_2/I_1) must satisfy

$$H'(s) = H \left(\frac{1}{s} \right) \tag{9.29}$$

As an example, for the filter N of Fig. 9.24(a), we have

$$H(s) = \frac{E_2}{E_1} = \frac{12}{2s^2 + 7s + 6} \tag{9.30}$$

$$Z_{11}(s) = \frac{E_1}{I_1} = \frac{2s^2 + 7s + 6}{s(2s + 1)} \tag{9.31}$$

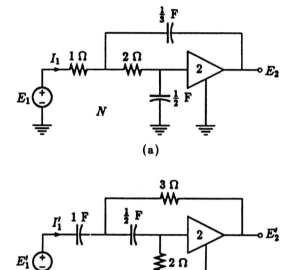

(a)

(b)

Figure 9.24: An example of RC-CR transformation.

The filter N' of Fig. 9.24(b) is obtained from N by the RC-CR transformation. Direct analysis will show that

$$H'(s) = \frac{E_2'}{E_1'} = \frac{12s^2}{6s^2 + 7s + 2} \tag{9.32}$$

$$Z_{11}' = \frac{E_1'}{I_1'} = \frac{6s^2 + 7s + 2}{s(s + 2)} \tag{9.33}$$

It is clear that the relationship between $H(s)$ and $H'(s)$ satisfies (9.29). The relationship between $Z_{11}(s)$ and $Z_{11}'(s)$ satisfies (9.27).

Equation (9.29) is important for active filter applications. It states that the RC-CR transformation transforms a lowpass filter into a highpass filter, and vice versa. Also this transformation transforms a bandpass filter into another bandpass filter.

9.8   Types of biquads

In the next three chapters, a great deal of attention will be paid to second-order filter sections, which are called biquads. The term 'biquad' was originally used to designate certain specific second-order circuit configurations. More recently, it has been used as a generic short name for 'second-order filter section,' because these sections, in general, have a network function that is the ratio of two quadratic polynomials.[1]

$$H(s) = \frac{a_2 s^2 + a_1 s + a_0}{s^2 + b_1 s + b_0} \tag{9.34}$$

As was described in Section 8.3, the denominator of (9.34) is often written as

$$s^2 + b_1 s + b_0 = s^2 + \left(\frac{\omega_0}{Q}\right) s + \omega_0^2 \tag{9.35}$$

where $Q = \sqrt{b_0}/b_1$ and $\omega_0 = \sqrt{b_0}$. We shall use whichever of the two forms of (9.35) happens to be more convenient in a particular situation.

There are several special biquads that are useful and considered standard in active filters. Their network functions and the general patterns of variation of their magnitudes are listed in Table 9.1. The following are some observations about each type of biquad.

(1) **Lowpass biquad.** The dc gain is normalized to be G. The magnitude characteristic generally decreases from G to 0 as ω increases. When Q is high, there will be a peak in the neighborhood of ω_0.

(2) **Highpass biquad.** The high-frequency gain is normalized to G. The magnitude characteristic may have a peak near ω_0 if Q is sufficiently high.

(3) **Bandpass biquad.** In this type of biquad, $a_0 = a_2 = 0$. This type of characteristic was discussed at length in Section 8.3.

(4) **Bandreject biquad.** In this type of biquad, $a_1 = 0$. Each biquad has a pair of transmission zeros at $\pm j\sqrt{a_0/a_2}$. Depending on the values

---

[1]Some of the numerator coefficients may be zero.

of the coefficients, the dc gain (a_0/b_0) and the high-frequency gain (a_2) may each assume any value. They can be equal, comparable, or drastically different. If the high-frequency gain is far less than the dc gain, then the section is a *lowpass bandreject biquad*. If the dc gain is far less than the high-frequency gain, then the section is a *highpass bandreject biquad*. A bandreject biquad is also known as a *notch biquad*.

(5) **Allpass biquad.** This type of biquad has no effect on the gain characteristics. Its chief use is to improve the phase or delay characteristics of filters. These sections are called delay-equalization or phase-linearization sections. Their application is explained in Section 3.6.

9.9 Summary

This chapter contains a collection of miscellaneous topics that are elementary to active filter studies. Some commonly used notations and terminology are defined and they will be used in active filters. Some of the basic ideas, definitions, circuits, and formulas will be used at various points in the development of active filters in subsequent chapters. Other topics, such as transmission zeros and gain adjustments, are common to many active filters and it's expedient for them to be treated together at the outset. These topics are brought together here to enable a smoother and more coherent presentation of the next four chapters.

Problems

9.1 Obtain the transfer impedance E_2/I_1 in terms of the resistances shown in the following circuit.

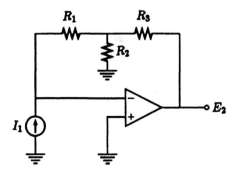

Table 9.1: **Standard types of biquads**

Types	Network functions	Magnitude characteristics
Lowpass	$H_{\mathrm{LP}}(s) = \pm\dfrac{Gb_0}{s^2+b_1s+b_0}$	
Highpass	$H_{\mathrm{HP}}(s) = \pm\dfrac{Gs^2}{s^2+b_1s+b_0}$	
Bandpass	$H_{\mathrm{BP}}(s) = \pm\dfrac{Gb_1s}{s^2+b_1s+b_0}$	
Bandreject	$H_{\mathrm{BR}}(s) = \pm\dfrac{a_2s^2 + a_0}{s^2+b_1s+b_0}$	
Allpass	$H_{\mathrm{AP}}(s) = \pm G\dfrac{s^2-b_1s+b_0}{s^2+b_1s+b_0}$	

9.2 Show that I is independent of the value of R_L. Obtain its value in terms of E_1, R_1, R_2, and R_3.

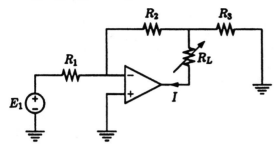

9.3 Show that the following circuit is a noninverting integrator.

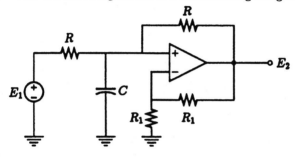

9.4 Give the element values in the following circuit so that it realizes the general first-order transfer function of (9.13). Let $R_1 = 1\,\Omega$.

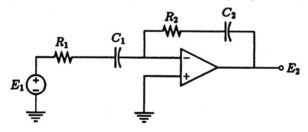

9.5 Give the element values in the following circuit so that it realizes the general first-order transfer function of (9.15). Let $R_1 = 1\,\Omega$.

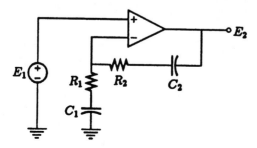

9.6 Realize the following voltage transfer functions by cascading two first-order sections.

(a) $-\dfrac{s+6}{(s+2)(s+4)}$

(b) $\dfrac{5}{s+8} + \dfrac{5}{s+10}$

(c) $\dfrac{12}{s^2+8s+12}$

9.7 Obtain the voltage transfer function of the following circuit.

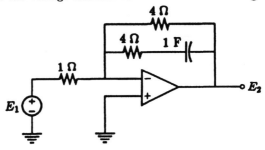

9.8 Show that the following circuit is a first-order allpass section.

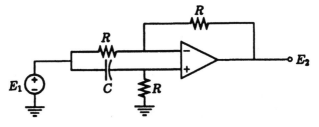

9.9 Find the simplest passive circuit that has the same impedance, Z_{in}, as the active circuit below.

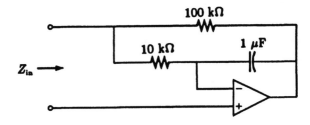

9.10 First obtain the parameters y_{31}, y_{32}, and y_{33} of the RC threeport below. Then use (9.19) to obtain the voltage transfer function.

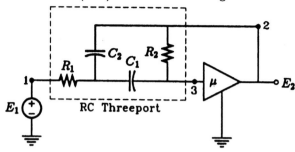

9.11 Derive the ratio E_2/E_1 in terms of the y parameters of the RC threeport. Under what condition will this circuit be stable?

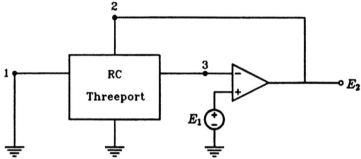

9.12 It is known that the following circuit realizes the voltage transfer function

$$\frac{E_2}{E_1} = -\frac{2}{s+1}$$

(a) Modify the circuit to reduce the gain to 1/1.5 of its original value. (b) Modify the circuit to increase the gain by a factor of 3 without using another op amp.

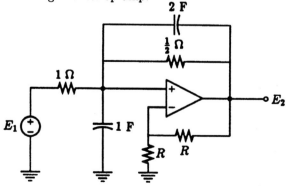

9.13 Without actually deriving the transfer function, find the two transmission zeros of the following circuit.

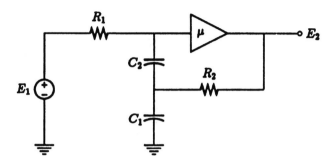

9.14 Without actually deriving the transfer function, state the form of the transfer function for the following circuit.

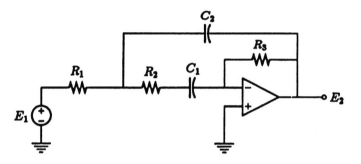

9.15 Obtain the voltage transfer function of the following circuit.

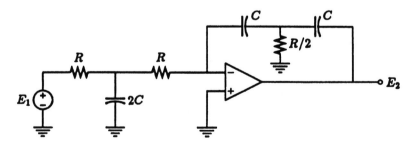

9.16 Network N is a second-order bandpass filter with zero output impedance and

$$\frac{E_2}{E_1} = \frac{2s}{s^2 + 5s + 3}$$

Obtain the voltage transfer function E_3/E_1.

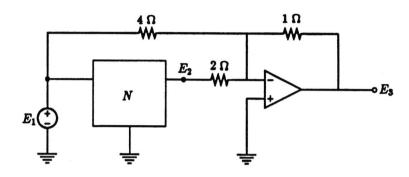

9.17 Derive the voltage transfer function of the following circuit.

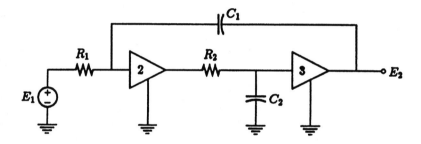

9.18 Derive the voltage transfer function of the following circuit.

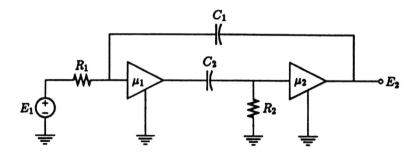

9.19 First obtain the voltage transfer function E_2/E_1 for the following circuit. Then determine the necessary value of μ in terms of R_1

and R_2 such that the circuit is an allpass biquad.

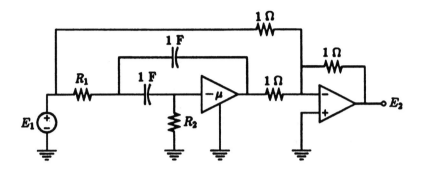

9.20 Derive the voltage ratio, E_2/E_1, in the following circuit in terms of R_1, R_2, and the y parameters of the RC threeport.

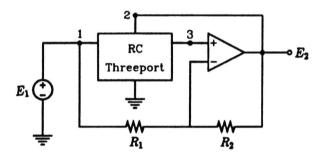

9.21 Derive the voltage ratio, E_2/E_1, of the following circuit.

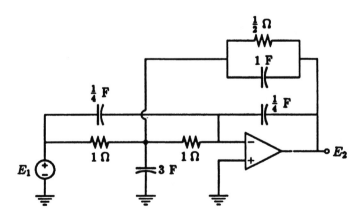

9.22 Derive the voltage ratio, E_2/E_1, of the following circuit. Simplify your answer.

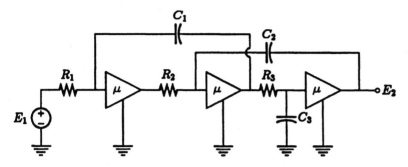

9.23 Derive the voltage ratio, E_2/E_1, of the following circuit in terms of the admittances.

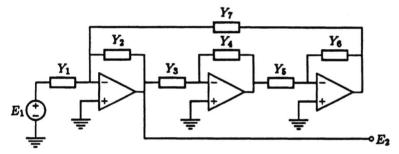

Chapter 10

Biquad circuits

In this chapter we shall present several widely used and well-established biquad circuits. The reader will find that the analysis and synthesis of these circuits are rather simple. Usually, given a circuit, it's a matter of writing a few node equations and obtaining the voltage transfer function. Then the coefficients of the desired transfer function are matched with the transfer function coefficients in terms of the circuit element values. Various related concepts and other basic techniques and formulas used here were developed and explained in Chapter 9. Sensitivities may be an issue; we already have developed the basic methods of evaluating them in Chapter 8.

The reader should also keep in mind that there are numerous other circuits available in the literature. What we include here is sufficient for most practical applications and, certainly, to serve as in introduction to the general area of active filters.

10.1 Sallen-Key biquads

Sallen and Key produced a list of active filters in 1954 [SK]. Many of these circuits are basic and have remained useful through the many decades since their publication. We shall study in detail several of the biquads from this group.

10.1.1 Lowpass biquad

The circuit of Fig. 10.1 is a lowpass biquad. To analyze this circuit, we shall use MATLAB. We write three node equations - one for the node whose voltage is E_3 and one each for the input terminals of the op amp (with the voltage designated as E_4). The following are the steps.

```
> f1 = '(E3-E1)/R1 + (E3-E4)/R2 + s*C1*(E3-E2)';
> f2 = '(E4-E3)/R2 + s*C2*E4';
> f3 = 'E4-E2/u';
> pretty(solve(f1,f2,f3,'E2,E3,E4'))
E2 =u E1/%1
                2
%1 := s C2 R2 + R2 s  C1 R1 C2 - u s C1 R1 + 1
+ R1 s C2 + s C1 R1
% Solutions for E3 and E4 have been discarded
```

Hence we have the voltage transfer function

$$\frac{E_2}{E_1} = \frac{\dfrac{\mu}{R_1 R_2 C_1 C_2}}{s^2 + \left(\dfrac{1}{R_1 C_1} + \dfrac{1}{R_2 C_1} + \dfrac{1-\mu}{R_2 C_2}\right)s + \dfrac{1}{R_1 R_2 C_1 C_2}} \qquad (10.1)$$

where $\mu = 1 + \frac{R_b}{R_a}$.

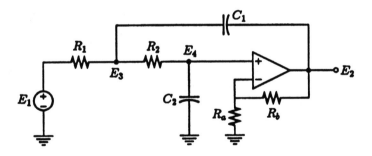

Figure 10.1: The Sallen-Key lowpass biquad.

Using the standard notation

$$H_{\mathrm{LP}}(s) = \frac{G\omega_0^2}{s^2 + \left(\dfrac{\omega_0}{Q}\right)s + \omega_0^2} \qquad (10.2)$$

we have

$$\omega_0 = \frac{1}{\sqrt{R_1 R_2 C_1 C_2}} \tag{10.3}$$

$$Q = \frac{\dfrac{1}{\sqrt{R_1 R_2 C_1 C_2}}}{\dfrac{1}{R_1 C_1} + \dfrac{1}{R_2 C_1} + \dfrac{1 - \mu}{R_2 C_2}} \tag{10.4}$$

$$G = \mu \tag{10.5}$$

If no restriction is imposed on the gain G, we have five element values to satisfy (10.3) and (10.4). Hence we are free to make some arbitrary choices. These arbitrary choices also offer us opportunities to take practical aspects of the circuit into consideration. The following examples of different choices of element values will illustrate how these opportunities can be utilized.

Design 1. *Equal element values.* In this design, we let

$$C_1 = C_2 = 1 \text{ F} \quad \text{and} \quad R_1 = R_2 = R \tag{10.6}$$

Equations (10.3) and (10.4) require

$$R = \frac{1}{\omega_0} \quad \text{and} \quad \mu = 3 - \frac{1}{Q} \tag{10.7}$$

The advantage of this design is that the calculation is simple, and the values of elements of the same type are uniform and easy to match.

Design 2. *Equal capacitance and equal feedback resistances.* In this design, we choose

$$C_1 = C_2 = 1 \text{ F} \quad \text{and} \quad R_a = R_b = R \tag{10.8}$$

Equations (10.3) and (10.4) require

$$R_1 = \frac{Q}{\omega_0} \quad \text{and} \quad R_2 = \frac{1}{\omega_0 Q} \tag{10.9}$$

The advantage of this design is that the resistors for the voltage amplifier are identical and easy to match. The amplifier gain is fixed at 2. The disadvantage is that the values of R_1 and R_2 can be very disparate as $R_1/R_2 = Q^2$.

Design 3. *Moderate-sensitivity design.* In this design, we choose

$$C_1 = \sqrt{3}Q \qquad C_2 = 1 \text{ F} \qquad \mu = \frac{4}{3} \tag{10.10}$$

which leads to

$$R_1 = \frac{1}{\omega_0 Q} \qquad \text{and} \qquad R_2 = \frac{1}{\sqrt{3}\omega_0} \tag{10.11}$$

This design was suggested by Saraga [Sa]. As will be shown further, this design renders a compromise in the sensitivities with respect to passive and active elements.

Design 4. *Minimum sensitivity.* In this design, we choose

$$\mu = 1 \qquad \text{and} \qquad R_1 = R_2 = 1 \ \Omega \tag{10.12}$$

In such a choice, the amplifier reduces to a voltage follower. It not only saves two resistors, but also makes the gain insensitive since no resistor is used in the negative feedback part of the circuit. This design requires

$$C_1 = \frac{2Q}{\omega_0} \qquad \text{and} \qquad C_2 = \frac{1}{2Q\omega_0} \tag{10.13}$$

As we shall see in the comparison of the sensitivities of these designs, most of the sensitivities of this design are either zero or extremely low. However, this design requires a very large spread of capacitance values as (10.13) gives $C_1/C_2 = 4Q^2$.

Sensitivity comparison

Using the formulas given in Chapter 8 for calculating the sensitivities, we obtain the following expressions:

$$S_{R_1}^{\omega_0} = S_{R_2}^{\omega_0} = S_{C_1}^{\omega_0} = S_{C_2}^{\omega_0} = -\frac{1}{2}$$

$$S_{\mu}^{\omega_0} = 0$$

$$S_{R_1}^{Q} = -\frac{1}{2} + Q\sqrt{\frac{R_2 C_2}{R_1 C_1}}$$

$$S_{R_2}^{Q} = -\frac{1}{2} + Q\left(\sqrt{\frac{R_1 C_2}{R_2 C_1}} + (1 - \mu)\sqrt{\frac{R_1 C_1}{R_2 C_2}}\right)$$

$$S_{C_1}^Q = -\frac{1}{2} + Q\left(\sqrt{\frac{R_1C_2}{R_2C_1}} + \sqrt{\frac{R_2}{C_2}R_1C_1}\right)$$

$$S_{C_2}^Q = -\frac{1}{2}(1-\mu)Q\sqrt{\frac{R_1C_1}{R_2C_2}}$$

$$S_{R_b}^Q = -S_{R_a}^Q = -(1-\mu)Q\sqrt{\frac{R_1C_1}{R_2C_2}}$$

$$S_\mu^Q = \mu Q\sqrt{\frac{R_1C_1}{R_2C_2}}$$

$$S_{R_1}^G = S_{R_2}^G = S_{C_1}^G = S_{C_2}^G = 0$$

$$S_\mu^G = 1$$

It is easily seen that $S_{R_1}^{\omega_0}$, $S_{R_2}^{\omega_0}$, $S_{C_1}^{\omega_0}$, $S_{C_2}^{\omega_0}$, $S_{R_1}^G$, $S_{R_2}^G$, $S_{C_1}^G$, and $S_{C_2}^G$ are all either zero or very low and they are independent of the design. Those sensitivities that are different for different designs are tabulated in Table 10.1.

From this tabulation, a trend can be observed. Design 1 is the simplest and easiest to implement in terms of element values. It also has the highest sensitivities. Design 2 is less sensitive than Design 1. This reduction in sensitivity is attained at the expense of the wide resistance value spread. Design 4 is the least sensitive of all four designs. This is achieved at the expense of large capacitance value spread. Design 3 is a compromise between Design 2 and Design 4. Its sensitivities lie somewhere between Design 2 and Design 4. The element value spread is shared by both the capacitances and the resistances.

EXAMPLE 10.1. Design a fourth-order Butterworth lowpass filter using the cascade of two Sallen-Key biquads using Designs 1 and 2 respectively. Then let $\omega_0 = 2\pi \times 1000$ rad/sec and use 0.1-μF capacitors.

SOLUTION From Table A.1, we have, for the normalized network function

$$\frac{E_2}{E_1} = \frac{G_1}{s^2 + 0.765367s + 1} \times \frac{G_2}{s^2 + 1.847759s + 1}$$

Use Design 1 for the first biquad. We have

Table 10.1: **Tabulation of sensitivities of the four designs**

	Design 1	Design 2	Design 3	Design 4
μ	$3 - \frac{1}{Q}$	2	$\frac{4}{3}$	1
$S_{R_1}^Q$	$-\frac{1}{2} + Q$	$-\frac{1}{2} + Q$	$-\frac{1}{2} + \frac{1}{\sqrt{3}}Q$	0
$S_{R_2}^Q$	$\frac{1}{2} - Q$	$\frac{1}{2} - Q$	$\frac{1}{2} - \frac{1}{\sqrt{3}}Q$	0
$S_{C_1}^Q$	$-\frac{1}{2} + 2Q$	$\frac{1}{2} + Q$	$\frac{1}{2} + \frac{1}{\sqrt{3}}Q$	$\frac{1}{2}$
$S_{C_2}^Q$	$\frac{1}{2} - 2Q$	$-\frac{1}{2} - Q$	$-\frac{1}{2} - \frac{1}{\sqrt{3}}Q$	$-\frac{1}{2}$
S_{μ}^Q	$3Q - 1$	$2Q$	$\frac{3}{3\sqrt{3}}Q$	$\frac{1}{2}*$
$S_{R_a}^Q$	$1 - 2Q$	-1	$-\frac{1}{\sqrt{3}}Q$	0
$S_{R_b}^Q$	$2Q - 1$	1	$\frac{1}{\sqrt{3}}$	0
$S_{R_a}^\mu$	$-\dfrac{2Q-1}{3Q-1}$	$-\frac{1}{2}$	$-\frac{1}{4}$	0
$S_{R_b}^\mu$	$\dfrac{2Q-1}{3Q-1}$	$\frac{1}{2}$	$\frac{1}{4}$	0

*Since a voltage follower is used, this figure is merely academic.

$$C_1 = C_2 = 1 \text{ F} \qquad R_1 = R_2 = 1 \text{ } \Omega$$

$$\mu = 3 - \frac{1}{Q} = 2.23463 = 1 + \frac{R_b}{R_a}$$

In order to minimize the dc offset voltage, we make the resistances between the two op amp terminals to ground to be identical, or

$$R_1 + R_2 = \frac{R_a R_b}{R_a + R_b}$$

Solving, we get

$$R_a = 3.61991 \text{ } \Omega \qquad \text{and} \qquad R_b = 4.46926 \text{ } \Omega$$

Next we use Design 2 on the second biquad. We have

$$C_1 = C_2 = 1 \text{ F} \qquad R_1 = \frac{Q}{\omega_0} = \frac{1}{b_1} = 0.541199 \text{ } \Omega$$

$$R_2 = \frac{1}{Q\omega_0} = 1.84775 \text{ } \Omega$$

Again, making $R_a || R_b = R_1 + R_2$, we obtain

$$R_a = R_b = 4.77787 \text{ } \Omega$$

To denormalize the circuit, we first frequency-scale the elements by $k_f = 2\pi \times 10^3$. This makes all capacitance values 159.154 μF. In order to use 0.1-μF capacitors, we impedance-scale all elements by $k_z = 159.154/0.1 = 1591.54$. The filter is shown in Fig. 10.2. The dc gain is $G = G_1 G_2 = 2.2346 \times 2 = 4.4693$.

10.1.2 Highpass biquad

If we apply the RC-CR transformation to the lowpass biquad of Fig. 10.1, we will obtain a highpass biquad. Hence the circuit in Fig. 10.3 is a highpass biquad. Although it is possible to follow the element-by-element replacement of the transform, it is easier to rederive the transfer function of the new circuit. For the circuit of Fig. 10.3, we obtain

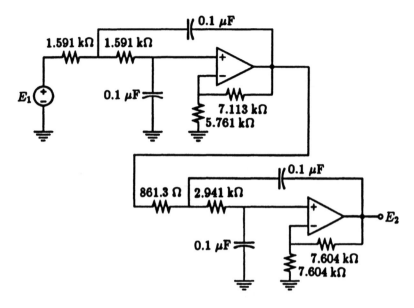

Figure 10.2: A fourth-order lowpass Butterworth filter using Sallen-Key biquads.

$$\frac{E_2}{E_1} = \frac{\mu s^2}{s^2 + \left(\dfrac{1}{R_2 C_2} + \dfrac{1}{R_2 C_1} + \dfrac{1-\mu}{R_1 C_1}\right)s + \dfrac{1}{R_1 R_2 C_1 C_2}} \tag{10.14}$$

Using the standard expression

$$H_{\text{HP}}(s) = \frac{Gs^2}{s^2 + \left(\dfrac{\omega_0}{Q}\right)s + \omega_0^2} \tag{10.15}$$

we can identify the following relationships:

$$\omega_0 = \frac{1}{\sqrt{R_1 R_2 C_1 C_2}} \tag{10.16}$$

$$Q = \frac{\dfrac{1}{\sqrt{R_1 R_2 C_1 C_2}}}{\dfrac{1}{R_2 C_2} + \dfrac{1}{R_2 C_1} + \dfrac{1-\mu}{R_1 C_1}} \tag{10.17}$$

$$G = \mu \tag{10.18}$$

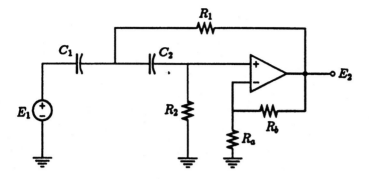

Figure 10.3: The Sallen-Key highpass biquad.

Similar to the lowpass biquad, if we do not put any restriction on G, there are five circuit elements whose values can be chosen in a wide variety of ways to satisfy Equations (10.16) and (10.17). We also have a great deal of flexibility in choosing those element values.

For example, let us choose $C_1 = C_2 = 1$ F. Then

$$\frac{1}{R_1 R_2} = b_0 \quad \text{and} \quad \frac{2}{R_2} + \frac{1-\mu}{R_1} = b_1$$

Solving these equations for R_1 and R_2, we obtain the relationships

$$R_2 = \frac{\sqrt{8b_0(\mu - 1) + b_1^2} - b_1}{2(\mu - 1)b_0} \quad \text{and} \quad R_1 = \frac{1}{b_0 R_2} \qquad (10.19)$$

10.1.3 Bandpass biquad

To realize a Sallen-Key bandpass biquad, one more element is required. One such circuit is shown in Fig. 10.4.

Analysis will yield

$$\frac{E_2}{E_1} = \frac{\dfrac{\mu}{R_1 C_1} s}{s^2 + \left(\dfrac{1}{R_1 C_1} + \dfrac{1}{R_3 C_2} + \dfrac{1}{R_3 C_1} + \dfrac{1-\mu}{R_2 C_1}\right)s + \dfrac{R_1 + R_2}{R_1 R_2 R_3 C_1 C_2}} \qquad (10.20)$$

Using the standard expression for a bandpass transfer function

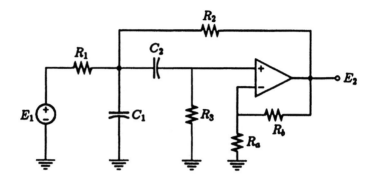

Figure 10.4: A Sallen-Key bandpass biquad.

$$H_{\mathrm{BP}}(s) = \frac{G\left(\dfrac{\omega_0}{Q}\right)s}{s^2 + \left(\dfrac{\omega_0}{Q}\right)s + \omega_0^2} \qquad (10.21)$$

we readily identify the following expressions for the three parameters:

$$\omega_0 = \sqrt{\frac{R_1 + R_2}{R_1 R_2 R_3 C_1 C_2}} \qquad (10.22)$$

$$Q = \frac{\sqrt{\dfrac{R_1 + R_2}{R_1 R_2 R_3 C_1 C_2}}}{\dfrac{1}{R_1 C_1} + \dfrac{1}{R_3 C_2} + \dfrac{1}{R_3 C_1} + \dfrac{1 - \mu}{R_2 C_1}} \qquad (10.23)$$

$$G = \frac{\dfrac{\mu}{R_1 C_1}}{\dfrac{1}{R_1 C_1} + \dfrac{1}{R_3 C_2} + \dfrac{1}{R_3 C_1} + \dfrac{1 - \mu}{R_2 C_1}} \qquad (10.24)$$

If there is no restriction on the value of G, there are six element values to satisfy two equations. Again, there is a great deal of flexibility in choosing the element values.

EXAMPLE 10.2. Use the equal-element criterion (Design 1) to obtain a bandpass biquad with $\omega_0 = 1$ and $Q = 10$.

SOLUTION Let $C_1 = C_2 = 1$ F and $R_1 = R_2 = R_3 = R$. From (10.22),

$$\omega_0 = \sqrt{\frac{2}{R^2}} \quad \Longrightarrow \quad R = \sqrt{2}\ \Omega$$

From (10.23),

$$\frac{\dfrac{\sqrt{2}}{R}}{\dfrac{3}{R} + \dfrac{1-\mu}{R}} = 10 \quad \Longrightarrow \quad \mu = 4 - \frac{\sqrt{2}}{10}$$

With this design, we obtain $G = 27.2843$.

In Chapter 9, we pointed out that if the RC-CR transformation is applied to a bandpass biquad, another bandpass biquad will result. The circuit of Fig. 10.5 is the circuit obtained from Fig. 10.4 by this procedure. This is another Sallen-Key bandpass biquad. Resistances R_a and R_b are not replaced by two capacitors[1] since their sole function is to form a voltage amplifier in conjunction with the op amp. This new circuit uses one more capacitor and is therefore less desirable.

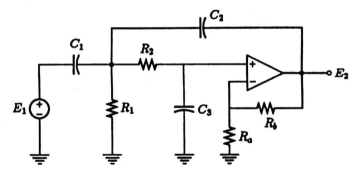

Figure 10.5: Another Sallen-Key bandpass biquad.

10.1.4 Bandreject biquad

The circuit of Fig. 10.6 uses the RC twin-tee network as the RC three-port. (See Fig. 9.15.) When the element values are properly chosen,

---

[1]In theory, we could replace these resistors with capacitors. But there is no good reason to do so. Discrete capacitors are more expensive. Also, using capacitors will leave no dc path to ground for the inverting terminal.

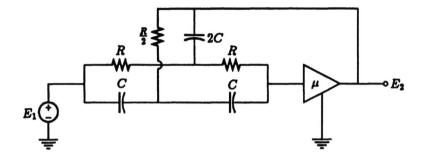

Figure 10.6: A Sallen-Key bandreject biquad.

the twin-tee network will produce a pair of j-axis transmission zeros. Analysis of the circuit will show that

$$\frac{E_2}{E_1} = \frac{\mu\left(s^2 + \dfrac{1}{R^2C^2}\right)}{s^2 + \dfrac{4(1-\mu)}{RC}s + \dfrac{1}{R^2C^2}} \tag{10.25}$$

which gives

$$Q = \frac{1}{4(1-\mu)} \tag{10.26}$$

Hence it is necessary that the gain μ be less than unity. This can be avoided by using a voltage follower and a slight change in the connection and the resistance values as shown in Fig. 10.7.

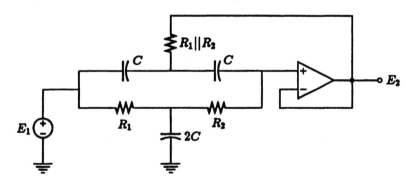

Figure 10.7: The modified biquad of Fig. 10.6 using a voltage follower.

The Sallen-Key bandreject biquad suffers from one shortcoming in that its dc and high-frequency gains are fixed to be the same. This biquad is usually used to eliminate a single frequency, such as a 120-Hz hum. It is not suitable for lowpass or highpass applications. However, a slight modification of the circuit will enable us to change either the dc or the high-frequency gain. For example, to reduce the high-frequency gain, we could apply the gain reduction technique described in Section 9.6.1 to the capacitor C connected to E_1. The modified circuit is shown in Fig. 10.8. Analysis will give

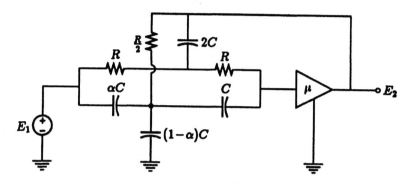

Figure 10.8: Modified lowpass Sallen-Key bandreject biquad.

$$\frac{E_2}{E_1} = \frac{\mu\alpha\left(s^2 + \dfrac{1}{\alpha R^2 C^2}\right)}{s^2 + \dfrac{4(1-\mu)}{RC}s + \dfrac{1}{R^2 C^2}} \tag{10.27}$$

A comparison of (10.27) with (10.25) will readily show that the denominator of the transfer function is unchanged. Hence the Q and ω_0 are unaffected. The dc gain remains at μ. The high-frequency gain is reduced by a factor α. At the same time, the transmission zeros are shifted from $\pm j1/RC$ to $\pm j1/\sqrt{\alpha}RC$.

If it is desired to reduce the dc gain, the same method can be applied to the resistor R connected to E_1. This will be left as an exercise.

As an alternative to the Sallen-Key circuit, a bandreject biquad can also be produced by the arrangement shown in the block diagram of Fig. 10.9. The input signal is fed forward and combined with the output of the bandpass filter. We then have

$$\frac{E_2}{E_1} = \frac{Gb_1 s}{s^2 + b_1 s + b_0} - G = -\frac{G(s^2 + b_0)}{s^2 + b_1 s + b_0} \tag{10.28}$$

which is the expression of a second-order notch filter.

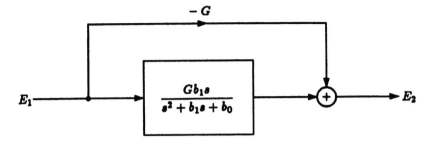

Figure 10.9: A bandreject biquad obtained by using a bandpass biquad.

10.2 Infinite-gain multiple-feedback (MFB) biquads

In this section, we shall study a class of biquads that use primarily a single op amp as an infinite-gain voltage amplifier in conjunction with an RC network. In the RC network, two or more elements provide negative feedback in the arrangement. This class of biquads shall be called the MFB biquads.

10.2.1 Lowpass biquad

The circuit of Fig. 10.10 has the voltage transfer function

$$\frac{E_2}{E_1} = -\frac{\dfrac{1}{R_1 R_3 C_1 C_2}}{s^2 + \dfrac{1}{C_1}\left(\dfrac{1}{R_1} + \dfrac{1}{R_2} + \dfrac{1}{R_3}\right)s + \dfrac{1}{R_2 R_3 C_1 C_2}} \qquad (10.29)$$

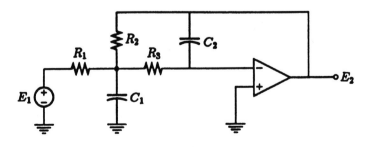

Figure 10.10: An MFB lowpass biquad.

This biquad is an inverting one and we shall use the standard form (Table 9.1) with the lower sign, namely,

$$\frac{E_2}{E_1} = -\frac{Gb_0}{s^2 + b_1 s + b_0} = -\frac{G\omega_0^2}{s^2 + \left(\frac{\omega_0}{Q}\right)s + \omega_0^2} \tag{10.30}$$

Hence

$$\omega_0 = \frac{1}{\sqrt{R_2 R_3 C_1 C_2}} \tag{10.31}$$

$$Q = \sqrt{\frac{C_1}{C_2}} \cdot \frac{1}{\dfrac{\sqrt{R_2 R_3}}{R_1} + \sqrt{\dfrac{R_3}{R_2}} + \sqrt{\dfrac{R_2}{R_3}}} \tag{10.32}$$

$$G = \frac{R_2}{R_1} \tag{10.33}$$

Now we have five elements and three equations. A great deal of flexibility is available for different designs. We arbitrarily choose $C_1 = 1$ F for convenience. Then we have

$$\frac{1}{R_1} + \frac{1}{R_2} + \frac{1}{R_3} = b_1 \qquad \frac{1}{R_2 R_3 C_2} = b_0 \qquad \frac{R_2}{R_1} = G$$

Solving these equations simultaneously, we get

$$R_2 = \frac{2(1 + G)}{b_1 + \sqrt{b_1^2 - 4C_2 b_0(1 + G)}} \tag{10.34}$$

$$R_1 = \frac{R_2}{G} \qquad R_3 = \frac{1}{b_0 R_2 C_2} \tag{10.35}$$

It is necessary to choose C_2 sufficiently small such that the quantity under the radical in (10.34) is positive.

EXAMPLE 10.3. Design a biquad using the circuit of Fig. 10.10 for $G = 5$, $b_1 = 1.2$, and $b_0 = 1$.

SOLUTION With $C_1 = 1$ F, we choose $C_2 = 0.05$ F. Then $R_2 = 7.1010\ \Omega$, $R_1 = 1.4202\ \Omega$, and $R_3 = 2.8165\ \Omega$.

From (10.32), it is seen that to achieve a high Q, it is necessary to make $C_2 \ll C_1$. Hence, high Q is achieved at the expense of wide capacitance value spread. On the other hand, this circuit has the advantage that the op amp is used as *the* amplifier. Since the op amp has extremely high gain, the biquad is insensitive to the amplifier gain. The other sensitivities are

$$S_{R_1}^{\omega_0} = 0$$

$$S_{R_2}^{\omega_0} = S_{R_3}^{\omega_0} = S_{C_1}^{\omega_0} = S_{C_2}^{\omega_0} = -\frac{1}{2}$$

$$S_{R_1}^{Q} = Q\sqrt{\frac{C_2}{C_1}}\left(\frac{\sqrt{R_2 R_3}}{R_1}\right)$$

$$S_{R_2}^{Q} = -\frac{Q}{2}\sqrt{\frac{C_2}{C_1}}\left(\frac{\sqrt{R_2 R_3}}{R_1} - \sqrt{\frac{R_3}{R_2}} + \sqrt{\frac{R_2}{R_3}}\right)$$

$$S_{R_3}^{Q} = -\frac{Q}{2}\sqrt{\frac{C_2}{C_1}}\left(\frac{\sqrt{R_2 R_3}}{R_1} + \sqrt{\frac{R_3}{R_2}} - \sqrt{\frac{R_2}{R_3}}\right)$$

$$S_{C_1}^{Q} = -S_{C_2}^{Q} = \frac{1}{2}$$

A comparison of the quantities in the parentheses in the above equations and the denominator of (10.32) shows that the former are all less than the latter. Hence we can infer that

$$|S_{R_1}^{Q}| < 1 \qquad |S_{R_2}^{Q}| < \frac{1}{2} \qquad |S_{R_3}^{Q}| < \frac{1}{2}$$

and this type of biquad typically has very low sensitivities.

10.2.2 Highpass biquad

Applying the RC-CR transformation to the circuit of Fig. 10.10, we obtain the highpass biquad circuit of Fig. 10.11. The voltage transfer function is

$$\frac{E_2}{E_1} = -\frac{\dfrac{C_1}{C_2}s^2}{s^2 + \dfrac{C_1 + C_2 + C_3}{R_2 C_2 C_3}s + \dfrac{1}{R_1 R_2 C_2 C_3}} \tag{10.36}$$

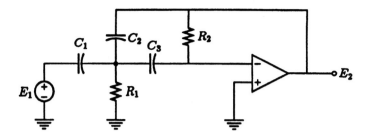

Figure 10.11: An MFB highpass biquad.

For design purposes, we can choose $C_1 = C_3 = 1$ F. The normalized circuit elements are given by

$$C_2 = \frac{1}{G} \qquad R_1 = \frac{Gb_1}{(2G+1)b_0} \qquad R_2 = \frac{2G+1}{b_1} \qquad (10.37)$$

10.2.3 Bandpass biquad

The circuit of Fig. 10.12 is an MFB bandpass biquad. The voltage transfer function can be obtained by analysis to be

$$\frac{E_2}{E_1} = -\frac{\dfrac{1}{R_1 C_2} s}{s^2 + \left(\dfrac{1}{R_2 C_1} + \dfrac{1}{R_2 C_2}\right) s + \dfrac{1}{R_1 R_2 C_1 C_2}} \qquad (10.38)$$

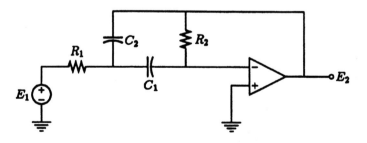

Figure 10.12: An MFB bandpass biquad.

Hence

$$\omega_0 = \frac{1}{\sqrt{R_1 R_2 C_1 C_2}} \qquad (10.39)$$

$$Q = \frac{\sqrt{\dfrac{R_2}{R_1}}}{\sqrt{\dfrac{C_2}{C_1}} + \sqrt{\dfrac{C_1}{C_2}}} \tag{10.40}$$

$$G = \frac{R_2 C_1}{R_1 (C_1 + C_2)} \tag{10.41}$$

To attain high Q it is desirable for the denominator of (10.40) to be minimum. It can be shown that the denominator is minimum when $C_1 = C_2$, which makes the denominator equal to 2. We thus choose $C_1 = C_2 = 1$ F, which leads to

$$Q = \frac{1}{2}\sqrt{\frac{R_2}{R_1}} \tag{10.42}$$

It is seen that to achieve high Q it is necessary to make $R_2 \gg R_1$. This disadvantage - high Q is achieved at the expense of wide element value spread - is again apparent.

With this choice of C_1 and C_2, we have

$$R_1 = \frac{b_1}{2b_0} = \frac{1}{2\omega_0 Q} \quad \text{and} \quad R_2 = \frac{2}{b_1} = \frac{2Q}{\omega_0} \tag{10.43}$$

and the gain attained is

$$G = \frac{R_2}{2R_1} = \frac{2b_0}{b_1^2} = 2Q^2 \tag{10.44}$$

If this gain is too high, we can apply the gain-reduction method described in Section 9.6.1 to R_1.

EXAMPLE 10.4. Design a bandpass biquad using the circuit of Fig. 10.12 such that $G = 1$, $\omega_0 = 1$, and $Q = 2$. Let $C_1 = C_2 = 1$ F.

SOLUTION   Using (10.43), we have

$$R_1 = \frac{1}{4} \ \Omega \quad \text{and} \quad R_2 = 4 \ \Omega$$

which renders $G = 8$. To reduce G to 1, we apply the method of Fig. 9.19 with (9.24) to R_1. The result is the circuit of Fig. 10.13.

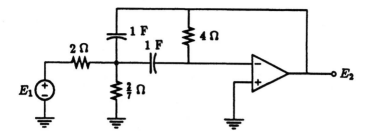

Figure 10.13: The circuit of Example 10.4.

A comparison between the voltage ratio of the MFB biquad (10.38) and that of the Sallen-Key biquad (10.20) bears out a salient difference. This difference is in the coefficients of the two middle terms of their denominators. In (10.38), all quantities in that coefficient are positive. In contrast, in (10.20), there is a negative term. Since these coefficients are inversely proportional to the Q's of their poles, to achieve high Q's, it is necessary to make these coefficients small. With the presence of the negative terms in (10.20) (as well as voltage ratios of other Sallen-Key biquads), it is relatively easy to achieve high Q's by adjusting μ. The only way to make this coefficient small in the MFB biquads is to make the passive element values large. Hence, generally speaking, the MFB biquads are suitable only for moderate- or low-Q applications.

Q enhancement

The Q of an infinite-gain MFB biquad can be enhanced by several methods. We shall describe two of these methods here. In the arrangement of Fig. 10.14, part of the output signal is fed back to combine with the input signal. This method requires that the gain of the biquad be unity. In Fig. 10.14, the following relationship must be satisfied:

$$E_2 = \left(\frac{-b_1 s}{s^2 + b_1 s + b_0} \right)(-E_1 - \beta E_2) \tag{10.45}$$

Rearranging, we get

$$\frac{E_2}{E_1} = \frac{b_1 s}{s^2 + (1 - \beta)b_1 s + b_0} \tag{10.46}$$

Thus Q_{new} is $Q/(1 - \beta)$. Hence, to increase Q we should choose β such that

$$\beta = 1 - \frac{Q}{Q_{\text{new}}} \tag{10.47}$$

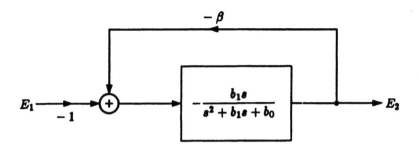

Figure 10.14: Q enhancement of a bandpass filter with unity gain.

As an example, the Q in the biquad of Fig. 10.13 can be enhanced by using the arrangement of Fig. 10.14. The actual circuit is shown in Fig. 10.15.

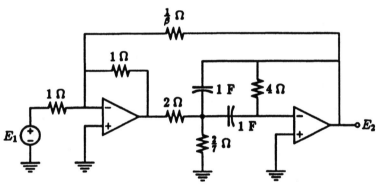

Figure 10.15: Enhancement of Q in the biquad of Fig. 10.13.

Another method that also employs positive feedback is shown in Fig. 10.16. Analysis of the circuit of Fig. 10.16 will yield

$$\frac{E_2}{E_1} = \frac{\dfrac{1}{R_1 C_2} s}{s^2 + \left(\dfrac{1}{R_2 C_1} + \dfrac{1}{R_2 C_2} - \dfrac{\beta}{R_1 C_2}\right) s + \dfrac{1 + \beta}{R_1 R_2 C_1 C_2}} \tag{10.48}$$

A comparison of (10.48) with (10.38) will show that for small values of β, the value of ω_0 is changed slightly. The value of b_1 can be greatly reduced if R_1 is less than R_2, thereby increasing Q substantially.

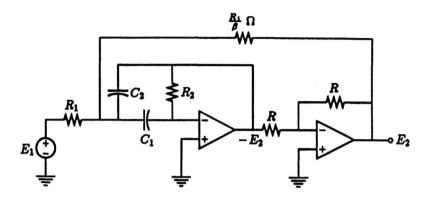

Figure 10.16: Another Q enhancement arrangement.

As can be seen in these two methods of Q enhancement, one common theme is to introduce a negative quantity in the coefficients of the middle term of the denominator. This makes b_1 small and increases Q.

10.2.4 Bandreject and allpass biquads

Bandreject and allpass biquads can be formed by using a bandpass biquad and a feed-forward arrangement similar to what was done using a Sallen-Key biquad shown in Fig. 10.9. The only difference is that the MFB biquads are inverting ones and the gain of the feed-forward path should have the opposite sign to that in Fig. 10.9.

However, for the MFB biquads, there is an alternative to the arrangement of Fig. 10.9. The feed-forward can also be applied to the noninverting terminal of the op amp as shown in Fig. 10.17.

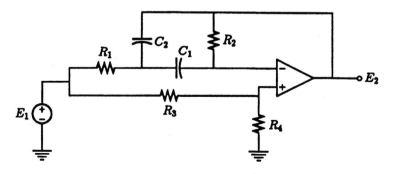

Figure 10.17: An MFB bandreject or allpass biquad.

Analysis of the circuit gives

$$\frac{E_2}{E_1} = \alpha \frac{s^2 + \left[\frac{1}{R_2 C_2} + \frac{1}{R_2 C_1} + \left(1 - \frac{1}{\alpha} \right) \frac{1}{R_1 C_2} \right] s + \frac{1}{R_1 R_2 C_1 C_2}}{s^2 + \left(\frac{1}{R_2 C_2} + \frac{1}{R_2 C_1} \right) s + \frac{1}{R_1 R_2 C_1 C_2}} \quad (10.49)$$

where $\alpha = R_4 / (R_3 + R_4)$. By choosing the value of α sufficiently small, we can make $a_1 = 0$ or $a_1 = -b_1$ and thus effect a bandreject or an allpass biquad respectively.

10.3 The two-integrator biquads

The development of this class of biquads can be related to the methods of realizing a system function using integrators. These realization procedures are well developed in system theory. The standard techniques of realizing system functions typically make use of the state-variable method of describing the system functions. Based on the state-variable description of the system, block diagrams are readily obtained.

For example, the block diagram of Fig. 10.18 can realize any second-order transfer function

$$\frac{E_2}{E_1} = G \frac{a_2 s^2 + a_1 s + a_0}{s^2 + b_1 s + b_0} \quad (10.50)$$

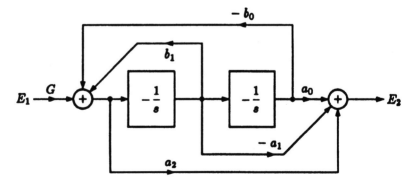

Figure 10.18: A standard realization of the transfer function of (10.50).

Several of the biquads in this section are based on this standard realization. Others are modifications of this or similar realizations.

10.3.1 The Kerwin-Huelsman-Newcomb (KHN) biquad

Kerwin, Huelsman, and Newcomb [KHN] first pointed out the application of the state-variable method of realizing a transfer function to active filters. The circuit of Fig. 10.19 consists of three interconnected op amp

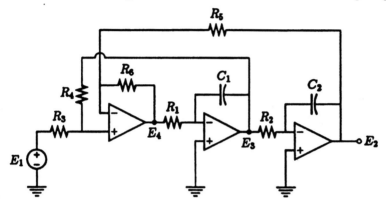

Figure 10.19: The KHN biquad.

circuits. The first op amp implements a differential summer. The other two are each used to implement an inverting integrator. For the two integrators, we have

$$E_2 = -\frac{1}{sR_2C_2}E_3 \tag{10.51}$$

$$E_3 = -\frac{1}{sR_1C_1}E_4 \tag{10.52}$$

For the differential summer,

$$E_4 = \frac{R_4(R_5 + R_6)}{R_5(R_3 + R_4)}E_1 - \frac{R_6}{R_5}E_2 + \frac{R_3(R_5 + R_6)}{R_5(R_3 + R_4)}E_3 \tag{10.53}$$

Combining (10.51), (10.52), and (10.53), we get

$$\frac{E_2}{E_1} = \frac{\dfrac{R_4}{R_5}\cdot\dfrac{R_5 + R_6}{R_3 + R_4}\cdot\dfrac{1}{R_1R_2C_1C_2}}{s^2 + \dfrac{R_3}{R_5}\cdot\dfrac{R_5 + R_6}{R_3 + R_4}\cdot\dfrac{1}{R_1C_1}s + \dfrac{R_6}{R_5}\cdot\dfrac{1}{R_1R_2C_1C_2}} \tag{10.54}$$

which gives

$$\omega_0 = \sqrt{\frac{R_6}{R_5} \cdot \frac{1}{R_1 R_2 C_1 C_2}} \tag{10.55}$$

$$Q = \frac{R_5}{R_3} \cdot \frac{R_3 + R_4}{R_5 + R_6} \sqrt{\frac{R_6}{R_5} \frac{R_1 C_1}{R_2 C_2}} \tag{10.56}$$

$$G = \frac{R_4}{R_6} \cdot \frac{R_5 + R_6}{R_3 + R_4} \tag{10.57}$$

The design formulas for such a biquad are very simple. The products $R_1 C_1$ and $R_2 C_2$ are the time constants of the two inverting integrators and can be chosen independently. Once these elements have been chosen, (10.55) requires

$$\frac{R_6}{R_5} = \omega_0^2 R_1 R_2 C_1 C_2 \tag{10.58}$$

With R_6/R_5 determined, (10.56) requires that

$$\frac{R_4}{R_3} = \frac{Q\left(1 + \dfrac{R_6}{R_5}\right)}{\sqrt{\dfrac{R_6}{R_5} \cdot \dfrac{R_1 C_1}{R_2 C_2}}} - 1 \tag{10.59}$$

EXAMPLE 10.5. Design a normalized KHN lowpass biquad with $\omega_0 = 1$ and $Q = 10$. Use $C_1 = C_2 = 1$ F and $R_1 = R_2 = 1\ \Omega$. Minimize the dc offset voltage. Calculate G.

SOLUTION   From (10.58), we choose $R_5 = R_6 = 1\ \Omega$. Then from (10.59), $\frac{R_4}{R_3} = 19$. To minimize the dc offset, we make $R_3 \| R_4 = R_5 \| R_6$, which requires $R_3 = \frac{10}{19}\ \Omega$ and $R_4 = 10\ \Omega$. From (10.57), $G = 1.9$.

Sensitivities

The sensitivities of ω_0, Q, and G can be derived by applying the formulas given in Chapter 8. The expressions are

$$S_{R_1}^{\omega_0} = S_{R_2}^{\omega_0} = S_{R_5}^{\omega_0} = -S_{R_6}^{\omega_0} = S_{C_1}^{\omega_0} = S_{C_2}^{\omega_0} = -\frac{1}{2}$$

$$S_{R_3}^{\omega_0} = S_{R_4}^{\omega_0} = 0$$

$$S_{R_1}^{Q} = -S_{R_2}^{Q} = S_{C_1}^{Q} = -S_{C_2}^{Q} = \frac{1}{2}$$

$$S_{R_3}^{Q} = -S_{R_4}^{Q} = -\frac{R_4}{R_3 + R_4}$$

$$S_{R_5}^{Q} = -S_{R_6}^{Q} = \frac{R_6 - R_5}{2(R_5 + R_6)}$$

$$S_{R_1}^{G} = S_{R_2}^{G} = 0$$

$$S_{R_3}^{G} = -S_{R_4}^{G} = -\frac{R_3}{R_3 + R_4}$$

$$S_{R_5}^{G} = -S_{R_6}^{G} = \frac{R_5}{R_5 + R_6}$$

These sensitivities are usually quite low. For the design in Example 10.5, those sensitivities that are not constant are

$$S_{R_3}^{Q} = -S_{R_4}^{Q} = -\frac{19}{20} \qquad S_{R_5}^{Q} = S_{R_6}^{Q} = 0$$

$$S_{R_3}^{G} = -S_{R_4}^{G} = -\frac{1}{20} \qquad S_{R_5}^{G} = -S_{R_6}^{G} = \frac{1}{2}$$

Highpass and bandpass outputs

Another very desirable feature of the KHN biquad is that both bandpass and highpass outputs are also available in the circuit. Thus this circuit can also be used for these two other types of biquads. If we take E_3 as the output, from (10.51), it is clear that E_3/E_1 is an inverting second-order bandpass function. If we take E_4 as the output, combining (10.51) and (10.52), we see that E_4/E_1 is a noninverting second-order highpass function. In fact, as was mentioned at the beginning of this chapter, with an additional summer, this biquad can be used to realize any general biquad. By properly summing E_4, E_3, and E_2 we are simply completing the op amp implementation of the block diagram of Fig. 10.18. Dependent on the signs of the numerator coefficients of (10.50), the summer may have to be of the differential type. Otherwise, an inverter can be inserted to render the proper signs of the contribution from each output.

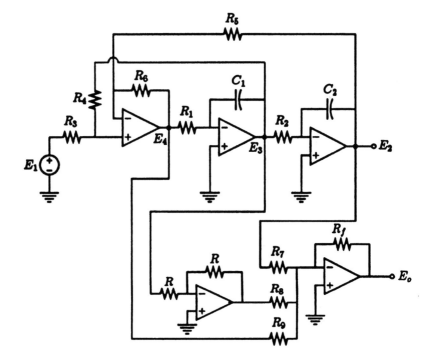

Figure 10.20: A general KHN biquad.

For example, if it is necessary that all numerator coefficients be positive, we can use the arrangement of Fig. 10.20.

Although the KHN biquad has low Q, ω_0, and G sensitivities, its magnitude sensitivities are not necessarily low. The feedback or feed-forward resistance values are directly related to the functional coefficients rather than the performance parameters. It usually requires differential summers or additional inverters in the circuits.

10.3.2   The Tow-Thomas biquad

In the arrangement of Fig. 10.18, the two feedback gains, b_1 and b_0, have opposite signs. This requires that the two feedback signals subtract. This subtraction can be avoided if an inverter is inserted either before or after the second integrator. At the same time, the input summer and the first integrator can be combined into a single summing integrator. These are the features of the Tow-Thomas biquad [Ths,To] shown in Fig. 10.21. Analysis yields

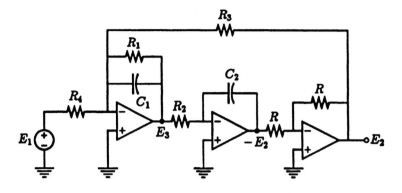

Figure 10.21: The Tow-Thomas biquad.

$$\frac{E_2}{E_1} = -\frac{\dfrac{1}{R_2 R_4 C_1 C_2}}{s^2 + \dfrac{1}{R_1 C_1}s + \dfrac{1}{R_2 R_3 C_1 C_2}} \tag{10.60}$$

and

$$\frac{E_3}{E_1} = -\frac{\dfrac{1}{R_4 C_1}s}{s^2 + \dfrac{1}{R_1 C_1}s + \dfrac{1}{R_2 R_3 C_1 C_2}} \tag{10.61}$$

Hence both the inverting and the noninverting lowpass output are available. So is the inverting bandpass output.

The design formulas for this biquad are also very simple. We can choose

$$C_1 = C_2 = 1 \text{ F} \qquad R_2 = R_3$$

Then for the lowpass case, we make

$$R_1 = \frac{1}{b_1} \qquad R_2 = R_3 = \frac{1}{\sqrt{b_0}} \qquad R_4 = \frac{R_3}{G} = \frac{1}{G\sqrt{b_0}}$$

For the bandpass case, we can use the same R_1, R_2, R_3, C_1, and C_2. We make

$$R_4 = \frac{1}{G b_1}$$

The Tow-Thomas biquad has the advantage that the noninverting terminals of all op amps are grounded. Its Q can be adjusted by varying a single resistance - R_1. The gain can be varied by varying a single resistance - R_4. Also the lowpass output can be either inverting or noninverting. It does not have a readily available highpass output. However, similar to the KHN biquad, by properly combining E_1, E_2, and E_3, any second-order transfer function can be obtained. The sensitivities of the Tow-Thomas biquad are generally comparable to those of the KHN biquad.

10.3.3   The Fleischer-Tow biquad

The Tow-Thomas biquad can be modified to realize any general second-order transfer function without using any additional op amp. Fleischer and Tow [FT] introduced two additional feed-forward paths to accomplish this. The modified circuit is shown in Fig. 10.22. Note also that the positions of the second integrator and the inverter have been interchanged.

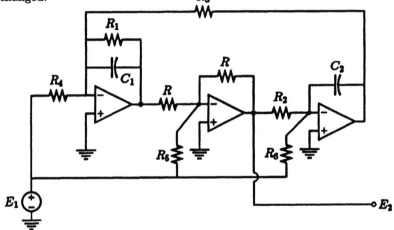

Figure 10.22: The Fleischer-Tow biquad.

Analysis of the circuit in Fig. 10.22 will give

$$\frac{E_2}{E_1} = -\frac{\dfrac{R}{R_5}s^2 + \dfrac{1}{R_1C_1}\left(\dfrac{R}{R_5} - \dfrac{R_1}{R_4}\right)s + \dfrac{1}{R_3R_6C_1C_2}}{s^2 + \dfrac{1}{R_1C_1}s + \dfrac{1}{R_2R_3C_1C_2}} \qquad (10.62)$$

It is seen that the denominator of (10.62) is identical to that of (10.61).

Hence, the values of R_1, R_2, R_3, C_1, and C_2 can be chosen in the same way as they were chosen for the Tow-Thomas biquad. Then the values of R_4, R_5, and R_6 can be chosen to satisfy any combination of numerator coefficients. For several special cases, the choice of these three resistances can be made as follows. (For this list, it is assumed that $C_1 = C_2 = 1$ F, $R_1 = 1/b_1$, $R_2 = R_3 = 1/\sqrt{b_0}$, and R is arbitrary.)

(1) Lowpass biquad.

$$R_4 = \infty \qquad R_5 = \infty \qquad R_6 = \frac{R_2}{G}$$

(2) Bandpass biquad.

$$R_5 = \infty \qquad R_6 = \infty \qquad R_4 = \frac{R_1}{G}$$

(3) Highpass biquad.

$$R_6 = \infty \qquad R_4 = \frac{R_1}{G} \qquad R_5 = \frac{R}{G}$$

(4) Bandreject biquad.

$$R_4 = \frac{R_5 R_1}{R}$$

R_5 and R_6 can be chosen to render different values of a_2 and a_0

(5) Allpass biquad.

$$R_5 = \frac{R}{G} \qquad R_6 = \frac{R_2}{G} \qquad R_4 = \frac{R_1}{2G}$$

10.4 Summary

In this chapter, several popular and practical biquads have been studied in detail. Their relative merits were pointed out as the developments were introduced. This compilation of different biquad circuits demonstrates the typical but happy dilemma of an active filter designer - that there are too many circuits from which to choose. There are situations in which certain biquads are clearly not suitable. An example of this is the use of MFB biquads in very high Q realizations. In other situations, several biquads may be equally suitable. Frequently, designers have to make a subjective choice as to which biquad to use for a particular task in hand based on factors other than quantitative comparison of several biquads.

The biquads presented in this chapter should be regarded as representative of all available biquads. Several additional biquads are included in the Problems section at the end of the chapter.

Problems

10.1 Design a Sallen-Key lowpass biquad with Chebyshev characteristic. Make $\omega_p = 1$ rad/sec and $\alpha_p = 0.5$ dB. Use Design 1. Then denormalize it so the new ω_p is 1000 rad/sec and the capacitors are 0.1 μF each.

10.2 Repeat Problem 10.1 using Saraga's design. Make the larger of the two capacitors 0.1 μF.

10.3 Design a Sallen-Key biquad to realize the voltage transfer function

$$\frac{Gs^2}{s^2 + 0.01s + 1}$$

Let $C_1 = C_2 = 1$ F and $R_1 = R_2$.

10.4 Design a Sallen-Key bandpass biquad with $\omega_0 = 2\pi \times 1000$ rad/sec, $Q = 10$, and $G = 4$. Use Design 1 and 0.01-μF capacitors.

10.5 We wish to reduce the dc gain of the Sallen-Key bandreject biquad by applying the method of Fig. 9.19 to the input resistor R of the circuit of Fig. 10.6. Describe its effect on the transmission zeros.

10.6 Design a Sallen-Key bandreject biquad to have the band center located at 60 Hz and $Q = 10$. Use a voltage follower as the active device.

10.7 The following circuit is a Sallen-Key lowpass biquad with a real zero. Obtain its voltage transfer function.

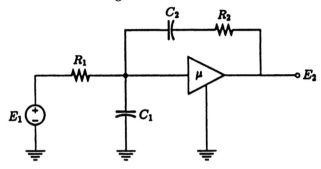

10.8 The following block diagram can be implemented using three op amps, two capacitors, and a number of resistors. Do this using 1-F capacitors and making the feedback resistor of the summing inverter 1 Ω. Also, obtain the voltage ratio, E_2/E_1, in terms of the gains.

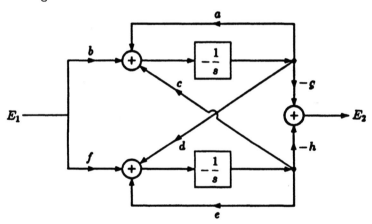

10.9 Design a normalized second-order Butterworth lowpass filter with $G = 2$, $C_1 = 1$ F, $C_2 = 0.005$ F using the MFB biquad of Fig. 10.10.

10.10 Design a normalized second-order Chebyshev lowpass filter with $\alpha_p = 2$ dB and $G = 2$. Use the MFB biquad with $C_1 = 1$ F and $C_2 = 0.05$ F.

10.11 Design a second-order bandpass MFB biquad with $\omega_0 = 2\pi \times 10^3$ rad/sec, $Q = 10$, and $G = 5$. Use $C_1 = C_2 = 0.1$ μF.

10.12 Calculate the value of β in Fig. 10.15 which increases the Q of the biquad by a factor of 10.

10.13 Positive feedback is introduced to enhance the Q of the biquad of Example 10.4 (before gain reduction) by the arrangement of Fig. 10.16. What will be the new values of ω_0 and Q if $\beta = 0.1$?

10.14 Derive the expressions for all ω_0, Q, and G sensitivities of the infinite-gain MFB bandpass biquad of Fig. 10.12.

10.15 Design a KHN lowpass biquad such that $\omega_0 = 1$ and $Q = 20$. Use $C_1 = C_2 = 1$ F and $R_1 = R_2 = 1\ \Omega$. Calculate G.

10.16 For the circuit obtained in the previous problem, calculate the sensitivities of ω_0, Q, and G with respect to all passive elements.

10.17 Design a Tow-Thomas lowpass biquad with $\omega_0 = 1$, $Q = 5$, and $G = 10$. Use $C_1 = C_2 = 1$ F.

10.18 Design a Tow-Thomas bandpass biquad with $\omega_0 = 1$, $Q = 15$, and $G = 5$. Use $C_1 = C_2 = 1$ F.

10.19 Design a bandreject biquad with $\omega_0 = \omega_z = 1$, $Q = 10$, and $H(0) = 1$. Use a Tow-Thomas biquad with $C_1 = C_2 = 1$ F.

10.20 The following circuit is known as the Åkerberg-Mossberg biquad [AM]. It is a variation of the Tow-Thomas biquad and has superior performance at high frequencies when nonideal op amps are used. Derive the voltage ratios E_3/E_1 and E_2/E_1.

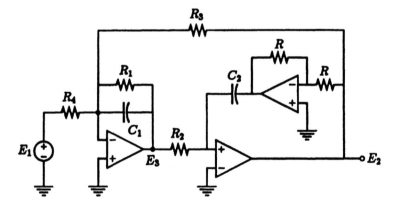

10.21 Design a Fleischer-Tow highpass biquad with $\omega_0 = 1$, $Q = 8$, and $G = 2$.

10.22 Design a Fleischer-Tow bandreject biquad with $\omega_0 = 1$, $Q = 12$, and $H(0) = H(\infty) = 1$.

10.23 Design a Fleischer-Tow allpass biquad with $\omega_0 = 1$, $Q = 5$, and $G = 1$.

10.24 A new bandpass biquad can be obtained by replacing the infinite-gain op amp of Fig. 10.12 with a finite-gain voltage amplifier of gain μ. Obtain ω_0, Q, and G in terms of the circuit element values and μ. (This circuit is known as the Delyiannis biquad [De].)

10.25 The following circuit is known as the Bainter bandreject biquad [Ba]. (a) Derive the transfer function E_2/E_1. (b) Choose $C_1 = C_2 = 1$ F. Develop a set of formulas that must be satisfied by the six resistances to satisfy the equality

$$\frac{E_2}{E_1} = \frac{s^2 + \omega_z^2}{s^2 + \left(\dfrac{\omega_0}{Q}\right)s + \omega_0^2}$$

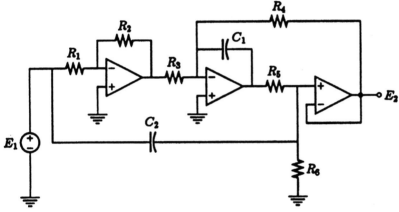

10.26 The following circuit is the Steffen allpass biquad [FHH]. (a) Derive the voltage ratio E_2/E_1. (b) Choose R_1, R_2, and R_3 to satisfy

$$\frac{E_2}{E_1} = \frac{s^2 - 5s + 10}{s^2 + 5s + 10}$$

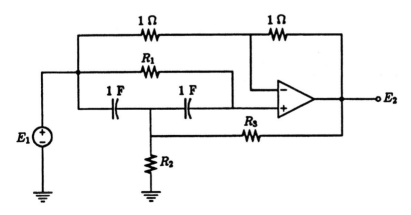

10.27 The following circuit is another version of the two-integrator bi-quad. Draw a block diagram for the circuit. Then obtain the transfer function, E_2/E_1.

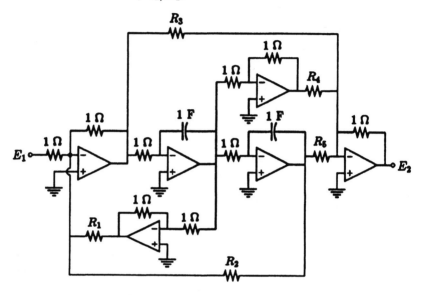

Chapter 11

High-order active filters

In Chapters 9 and 10, we laid the foundation for general, high-order active filters. In Chapter 9, several basic principles and useful tools were developed. In Chapter 10, we concentrated on the various second-order filter sections - the biquads. We considered several design criteria and formulas for obtaining element values of some better-known biquad circuits.

Now we are in a position to make use of the facilities developed in those two chapters in the design of high-order active filters. In this chapter, we will deal primarily with methods that make use of building blocks developed in the two previous chapters. In the following chapter, we will deal with techniques that rely on the active simulation of passive filters.

11.1 The cascade realization

As was mentioned earlier, the simplest way to realize a high-order filter function is to factor both the numerator and denominator polynomials and then form a number of fractional factors each of which is either a biquadratic or a bilinear function. Usually, only one bilinear function will be used and only when the order of the filter function is odd. In mathematical terms, we express a network function[1]

$$H(s) = \frac{a_m s^m + a_{m-1} s^{m-1} + \cdots + a_1 s + a_0}{s^n + b_{n-1} s^{n-1} + \cdots + b_1 s + b_0} \tag{11.1}$$

---

[1]This representation is entirely general. Frequently, many of the numerator coefficients are zero.

in the form of several biquadratic factors, plus one first-order factor if n is odd. Namely,

$$H(s) = \prod_{i=1}^{n/2} \frac{a_{2i}s^2 + a_{1i}s + a_{0i}}{s^2 + b_{1i}s + b_{0i}} = \prod_{i=1}^{n/2} H_i(s) \qquad n \text{ even} \qquad (11.2)$$

$$H(s) = \frac{a_{11}s + a_{01}}{s + b_{01}} \prod_{i=2}^{(n+1)/2} \frac{a_{2i}s^2 + a_{1i}s + a_{0i}}{s^2 + b_{1i}s + b_{0i}}$$

$$= H_1(s) \prod_{i=2}^{(n+1)/2} H_i(s) \qquad n \text{ odd} \qquad (11.3)$$

Since we have a fairly good repertoire of biquads and the bilinear functions are quite easy to realize (see Section 9.4), the realization of a general filter function is a matter of realizing a number of biquads (with possibly one additional first-order section) and then connecting them in cascade. We shall first illustrate this approach with a few examples.

EXAMPLE 11.1. Design a fifth-order lowpass Butterworth filter with a dc gain equal to unity and the half-power frequency at 1 kHz. Make the largest capacitance 0.1 μF.

SOLUTION   From Table A.1, the normalized transfer function is

$$H(s) = \frac{1}{s^2 + 0.618034s + 1} \cdot \frac{1}{s^2 + 1.61803s + 1} \cdot \frac{1}{s + 1}$$

$$= H_1(s) \cdot H_2(s) \cdot H_3(s) \qquad (11.4)$$

The variations of $|H_1(j\omega)|$, $|H_2(j\omega)|$, and $|H_3(j\omega)|$ are shown in Fig. 11.1, in which the magnitude of $H(j\omega)$ is also shown.

Since these pole Q's are quite low, the MFB biquads should be quite suitable for realizing $H_1(s)$ and $H_2(s)$. We shall use the biquad circuit of Fig. 10.10 for these two functions.

For $H_1(s)$ with $C_1 = 1$ F and $G = 1$ and from (10.34),

$$R_2 = \frac{2 \times 2}{0.618034 + \sqrt{0.618034^2 - 4 \times C_2 \times 2}}$$

Choose $C_2 = 0.02$ F and obtain

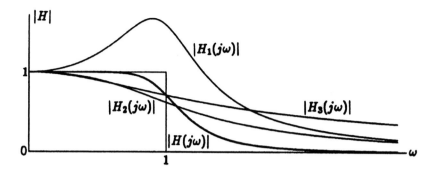

Figure 11.1: Magnitudes of $H(j\omega)$, $H_1(j\omega)$, $H_2(j\omega)$, and $H_3(j\omega)$ in (11.4).

$$R_2 = 3.6725 \ \Omega$$

Using (10.35), we get

$$R_1 = 3.6725 \ \Omega \quad \text{and} \quad R_3 = \frac{1}{b_0 R_2 C_2} = 13.615 \ \Omega$$

Similarly, for $H_2(s)$ with $C_1 = 1$ F and $G = 1$

$$R_2 = \frac{2 \times 2}{1.61803 + \sqrt{1.61803^2 - 4 \times C_2 \times 2}}$$

Choose $C_2 = 0.1$ F and obtain

$$R_2 = 1.3484 \ \Omega = R_1 \quad \text{and} \quad R_3 = \frac{1}{R_2 C_2 b_0} = 7.4159 \ \Omega$$

For $H_3(s)$, we use the first-order section of Fig. 9.13 with $a_1 = 0$ and $a_0 = b_0 = 1$. The normalized filter circuit is shown in Fig. 11.2. Since all three sections realize their respective transfer function with a negative sign, the circuit of Fig. 11.2 realizes the transfer function of (11.4) with a negative sign.

To denormalize the filter, we first frequency-scale the capacitors by $k_f = 2\pi \times 10^3$. That makes the 1-F capacitance 159.15 μF, the 0.02-F one 3.1831 μF, and the 0.1-F one 15.915 μF.

To scale the 159.15-μF capacitors to 0.1 μF, we impedance-scale the circuit by $k_z = 1591.5$. The final circuit is shown in Fig. 11.3.

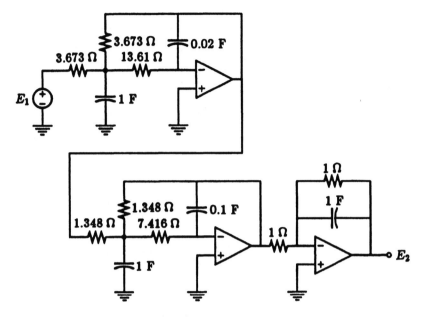

Figure 11.2: The normalized filter circuit of Example 11.1.

It should be pointed out that in the cascade realization, each section can stand alone and is an entity by itself. There need not be any commonality in the design choices among the different sections. In fact, each section should be designed entirely independently of all other sections. In Example 11.1, we have chosen each starting capacitance to be 1 F purely out of convenience. We could have started the design of each section with a different capacitance. Likewise, we could have impedance-scaled each section by a different k_z if we had had a good reason to do so.

From Example 11.1, it is seen that in realizing $H_1(s)$, even for a modest pole Q (1.62), the ratio of C_1 to C_2 is 50 to 1. This wide element spread is not unacceptable, but it is certainly not desirable. An inspection of the expression leading to the value of R_2 will reveal that if G is lowered, a larger value of C_2 can be used. This loss in gain can easily be made up by raising the gains of the other sections.

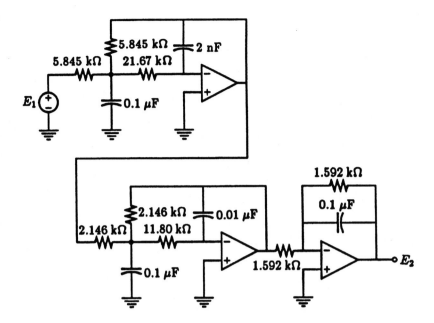

Figure 11.3: The denormalized filter circuit of Example 11.1.

EXAMPLE 11.2. Redesign the normalized filter of Example 11.1 so the highest capacitance spread in each section does not exceed 15 to 1.

SOLUTION In Example 11.1, the two capacitances in the biquad that realizes $H_1(s)$ have a ratio of 50 to 1. This choice was necessary to keep the quantity under the radical in the formula for R_2 positive. In order to increase the size of C_2, we could reduce the gain of $H_1(s)$. So we factor the same transfer function in (11.4) as

$$H(s) = \frac{0.2}{s^2 + 0.618034s + 1} \cdot \frac{1}{s^2 + 1.61803s + 1} \cdot \frac{5}{s + 1}$$

$$= H_1(s) \cdot H_2(s) \cdot H_3(s) \tag{11.5}$$

For the new $H_1(s)$ with $C_1 = 1$ F and $G = 0.2$, we have

$$R_2 = \frac{2 \times 1.2}{0.618034 + \sqrt{0.618034^2 - 4 \times C_2 \times 1.2}}$$

Choose $C_2 = \frac{1}{15}$ F, which gives

$$R_2 = 2.76828 \ \Omega \qquad R_1 = \frac{R_2}{G} = 13.8414 \ \Omega$$

$$R_3 = \frac{1}{R_2 C_2 b_0} = 5.41853 \ \Omega$$

Function $H_2(s)$ is unchanged from Example 11.1. So the same biquad can be used. The gain of the first-order section can be increased by a factor of 5 simply by reducing the input resistance by that factor. The new normalized filter circuit is shown in Fig. 11.4.

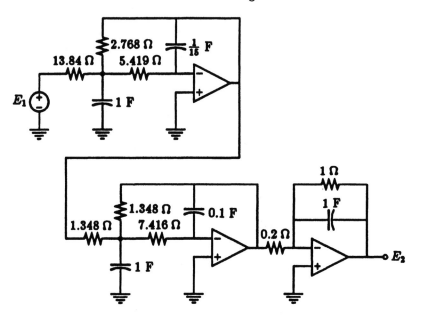

Figure 11.4: Another normalized filter circuit.

Gain allocation

The flexibility of allocating different gains to different filter sections gives an extra degree of freedom in designing high-order filters. This freedom of allocating different gains to different factors is known as *gain allocation* or *gain assignment*. Although it is difficult to give a general rule as to how the gain of each section in a cascaded filter should be, sometimes this flexibility offers an opportunity to improve the overall design.

Other than the element value spread consideration, as illustrated in Examples 11.1 and 11.2, different gain allocations will directly affect the signal strengths in each section. Hence, gain allocation can be used to make the maximum and minimum signal strengths in the pass band in all sections as uniform as practical. If this is achieved, the burden of attaining the overall gain is shared by all sections without making the signal in one particular section either too strong to cause distortion or op amp latching, or too weak that it falls below the noise threshold.

EXAMPLE 11.3. Design a normalized fourth-order bandpass Chebyshev filter with $\alpha_p = 1$ dB, $\Omega_0 = 1$, and $B = 0.1$.

SOLUTION From Table A.8, the lowpass prototype transfer function is

$$H_{\mathrm{LP}}(s) = \frac{K}{s^2 + 1.097734s + 1.102510}$$

We use the lowpass-to-bandpass transformation of (4.4) and let

$$s = \frac{S^2 + 1}{0.1S}$$

The result is[2]

$$H_{\mathrm{BP}}(S) = \frac{0.01KS^2}{S^4 + 0.109773S^3 + 2.01103S^2 + 0.109773S + 1} \tag{11.6}$$

At this point, we need to factor the denominator into two quadratic polynomials. We can use MATLAB for this purpose. Or else, the denominator zeros can be obtained from those of the lowpass prototype. The poles of the lowpass transfer function are $-0.548867 \pm j0.895129$. We solve two quadratic equations

$$S^2 - 0.1 \times (-0.548867 + j0.895129)S + 1 = 0$$

$$S^2 - 0.1 \times (-0.548867 - j0.895129)S + 1 = 0$$

to obtain the four poles of the bandpass transfer function. They are

$$-0.0286709 \pm j1.0453820 \quad \text{and} \quad -0.0262158 \pm j0.955869$$

Each pair of conjugate poles can now be paired to form one quadratic factor in the denominator of $H_{\mathrm{BP}}(S)$. We can now write

---

[2]We could use the MATLAB command 1p2bp as described in Section 4.2.

$$H_{\text{BP}}(S) = \frac{G_1 S}{S^2 + 0.057342 S + 1.09365}$$

$$\times \frac{G_2 S}{S^2 + 0.0524317 S + 0.914373} = H_1(S) \cdot H_2(S) \qquad (11.7)$$

Since the pole Q's are in the neighborhood of 20, we shall use the Sallen-Key biquads of Fig. 10.4. In each biquad, we choose $C_1 = C_2 = 1$ F and $R_1 = R_2 = R_3 = R$.

For $H_1(S)$ and from (10.21), we have

$$R = \frac{\sqrt{2}}{\omega_0} = \frac{\sqrt{2}}{\sqrt{1.09365}} = 1.3523 \ \Omega$$

From (10.20), it is necessary that

$$\frac{3}{R} + \frac{1 - \mu}{R} = 0.057342 \qquad \Longrightarrow \qquad \mu = 3.9225$$

and

$$G_1 = \frac{\mu}{R b_1} = 50.584$$

Making $R_a \| R_b = R_3$, we obtain $R_a = 1.8150 \ \Omega$ and $R_b = 5.3044 \ \Omega$. (The notation $R_a \| R_b$ represents the parallel combination of R_a and R_b.)

Similarly, for $H_2(S)$,

$$R = \frac{\sqrt{2}}{\omega_0} = \frac{\sqrt{2}}{\sqrt{0.914373}} = 1.4790 \ \Omega$$

and

$$\frac{3}{R} + \frac{1 - \mu}{R} = 0.052432 \qquad \Longrightarrow \qquad \mu = 3.9225$$

Also,

$$G_2 = \frac{\mu}{R b_1} = 50.584$$

Making $R_a \| R_b = R_3$, we obtain $R_a = 1.9850 \ \Omega$ and $R_b = 5.8011 \ \Omega$. The normalized filter circuit is shown in Fig. 11.5.

EXAMPLE 11.4. Design a normalized fourth-order lowpass elliptic filter with $\alpha_p = 1$ dB, $\omega_p = 1$, $\omega_s = 1.5$, and $\alpha_s = 39.52$ dB.

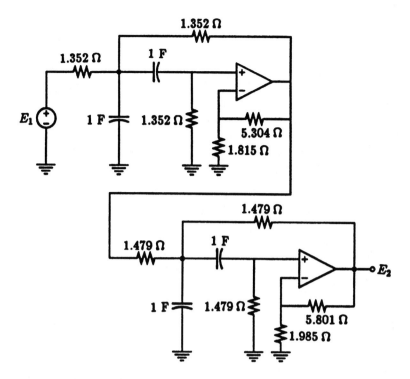

Figure 11.5: A normalized fourth-order Chebyshev bandpass filter.

SOLUTION From Table A.18, we have

$$H(s) = \frac{G_1(s^2 + 2.53555)}{s^2 + 0.208819s + 0.998811}$$

$$\times \frac{G_2(s^2 + 12.0993)}{s^2 + 0.729977s + 0.364281} = H_1(s) \cdot H_2(s) \qquad (11.8)$$

Since the filter calls for two lowpass notch biquads, we shall use the Fleischer-Tow biquad circuits of Fig. 10.22. In each biquad, we choose $C_1 = C_2 = 1$ F and $R_5 = R_6 = 1$ Ω.

For $H_1(s)$, we have

$$R_1 = \frac{1}{0.208819} = 4.7888 \ \Omega \qquad R = R_5 = 1 \ \Omega$$

$$R_3 = \frac{1}{a_0 R_6} = 0.39439 \ \Omega \qquad R_4 = \frac{R_1 R_5}{R} = 4.7888 \ \Omega$$

$$\frac{1}{R_2 R_3} = 0.99881 \quad \Longrightarrow \quad R_2 = 2.53858 \ \Omega$$

For $H_2(s)$,

$$R_1 = \frac{1}{0.729977} = 1.3699 \ \Omega \qquad\qquad R = R_5 = 1 \ \Omega$$

$$R_3 = \frac{1}{a_0 R_6} = 0.082649 \ \Omega \qquad\qquad R_4 = \frac{R_1 R_5}{R} = 1.3699 \ \Omega$$

$$\frac{1}{R_2 R_3} = 0.364281 \quad \Longrightarrow \quad R_2 = 33.2142 \ \Omega$$

The final filter circuit is shown in Fig. 11.6.

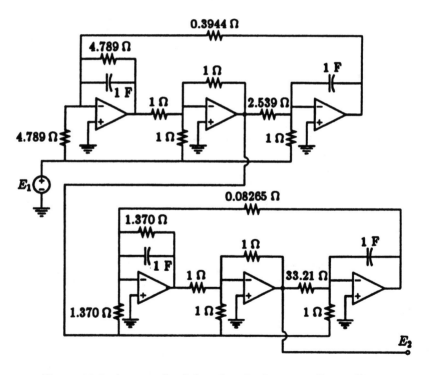

Figure 11.6: A normalized fourth-order lowpass elliptic filter.

Pole-zero pairing

In Example 11.3, we have written the transfer function as the product of two factors each of which is a bandpass transfer function. This is not the only way a function can be factored. We could rewrite the $H_{BP}(S)$ of (11.7) as

$$H_{BP}(S) = \frac{G_1 S^2}{S^2 + 0.057342S + 1.09365}$$

$$\times \frac{G_2}{S^2 + 0.0524317S + 0.914373} \qquad (11.9)$$

or

$$H_{BP}(S) = \frac{G_1}{S^2 + 0.057342S + 1.09365}$$

$$\times \frac{G_2 S^2}{S^2 + 0.0524317S + 0.914373} \qquad (11.10)$$

In each case, we would realize one highpass biquad and one lowpass biquad instead of two bandpass biquads as was done in the example.

Similarly, in Example 11.4, we could have regrouped the poles and zeros and let

$$H_1(s) = \frac{G_1(s^2 + 2.53555)}{s^2 + 0.729977s + 0.364281} \qquad (11.11)$$

and

$$H_2(s) = \frac{G_2(s^2 + 12.0993)}{s^2 + 0.208819s + 0.998811} \qquad (11.12)$$

instead of the way $H_1(s)$ and $H_2(s)$ were defined in (11.8).

This flexibility in the grouping poles and zeros in the various factors is known as *pole-zero pairing*. Again, it is difficult to give a simple general rule to determine which zeros should be associated with which pairs of poles. One technique for dealing with this issue soundly is to make an exhaustive study of all possible combinations and evaluate the ripples in the pass band in each section. Then the one whose maximum ripple in the pass band is minimum will be considered the best pairing.

Section sequencing

Another flexibility of the cascade realization is known as *section sequencing*. Since the transfer functions of all sections are eventually multiplied, mathematically it makes no difference how the sections are ordered. If there are three sections, as in Example 11.1, there are six different ways the sections can be cascaded.[3] They are

$$H_1 H_2 H_3 \quad H_1 H_3 H_2 \quad H_2 H_1 H_3 \quad H_2 H_3 H_1 \quad H_3 H_1 H_2 \quad H_3 H_2 H_1 \qquad (11.13)$$

Usually, for a moderately complex filter, this problem is not a critical one and some choices can be judiciously made by the engineer. For example, one recommended choice would be to place the lowpass section (if available) at the beginning of the cascade. This will prevent signals of unwanted frequencies from attaining sufficient strength before they reach the later sections. Also, it may be a good idea to place the highpass section (if available) at the end of the cascade. This will help to prevent the internally generated noise signal outside the pass band from reaching the output.

A quantitative comparison can also be conducted by an exhaustive study of the gains from the input to the outputs at all intermediate junctions for differently sequenced combinations. This is illustrated for the case of the three-section realization represented in Fig. 11.7. We then inves-

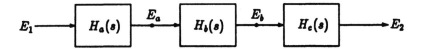

Figure 11.7: The cascade of three filter sections.

tigate the gains E_a/E_1 and E_b/E_1 for all six possible sequences listed in (11.13). The sequence that gives the flattest variations of these intermediate transfer functions can be considered to be the most desirable choice.

11.2   The state-variable method of realization

In this section, we shall describe several methods by which a filter transfer function can be realized directly without decomposing it into several

---

[3]If there are N sections, there are $N!$ different possible sequences.

lower-order factors. Most of the techniques used here either have identical counterparts or are closely related to realization techniques in system theory. Given a transfer function, a number of state variables are defined. The state matrices can then be written in terms of the coefficients of the transfer function. These techniques are commonly referred to as the *state-variable method* of realization.

11.2.1 Realization of all-pole transfer functions

It is well known in system theory that the block diagram of Fig. 11.8 realizes a general nth-order all-pole transfer function. This system uses n integrators and is known as the *controller canonic form* [?]. Equating the inputs of the summer to its output, we get

$$s^n E_2 = a_0 E_1 - b_{n-1} s^{n-1} E_2 - b_{n-2} s^{n-2} E_2 - \cdots - b_1 s E_2 - b_0 E_2 \qquad (11.14)$$

Solving gives

$$\frac{E_2}{E_1} = \frac{a_0}{s^n + b_{n-1} s^{n-1} + b_{n-2} s^{n-2} + \cdots + b_1 s + b_0} \qquad (11.15)$$

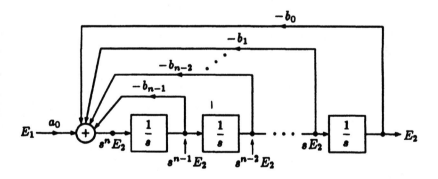

Figure 11.8: The controller canonic form of realization.

The system of Fig. 11.8 uses noninverting integrators, which are less suitable for op amp implementation than the inverting ones. The system can easily be modified by replacing every noninverting integrator with an inverting one and, at the same time, changing the sign of every other feedback gain as shown in Fig. 11.9.[4] This modification will change the sign of the overall gain when n is odd.

---

[4]The signs of b_1 and b_0 will depend on whether n is even or odd. Starting with b_{n-1}, the signs of the successive b's alternate.

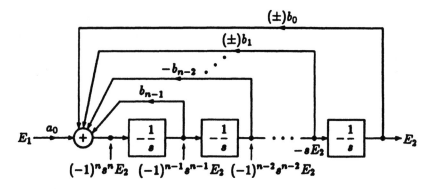

Figure 11.9: Realization of (11.15) using inverting integrators.

In implementing the block diagram of Fig. 11.9, it is not necessary to use an inverter for each feedback path whose gain is negative. Since eventually these feedback gains are all added in the summer, they can be summed and then the sum is inverted and fed to the input summer. Also, the summer and the first inverting integrator can be combined and replaced with an inverting weighted summing integrator. These implementations are illustrated for $n = 6$ in Fig. 11.10.

EXAMPLE 11.5. Realize a normalized fourth-order Chebyshev low-pass filter with $\omega_p = 1$, $\alpha_p = 1$ dB, and a dc gain of unity.

SOLUTION   From Table A.8, we have

$$\frac{E_2}{E_1} = \frac{0.275628}{s^4 + 0.952811s^3 + 1.453925s^2 + 0.742619s + 0.275628} \quad (11.16)$$

Following the procedure described, we obtain the circuit of Fig. 11.11.

11.2.2   Realization of general transfer functions

In the block diagram of Fig. 11.9, outputs $(-1)^i s^i E_2$, $i = 1, 2, \ldots, n-1$, are all available. (The output $(-1)^n s^n E_2$ has been absorbed into the weighted summing integrator and is no longer available.) By weighting and combining these outputs properly, any general nth-order transfer function for which $m < n$ can be readily obtained. It is easily seen that the arrangement of Fig. 11.12 realizes the transfer function of (11.1) as long as $m < n$.[5]

---

[5]The signs of a_{n-1} and a_{n-2} will depend on whether n is even or odd.

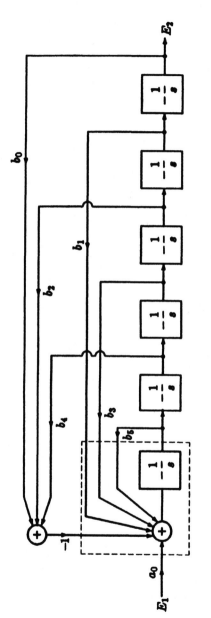

Figure11.10: Implementation of the system of Fig. 11.9 using one inverting weighted summer.

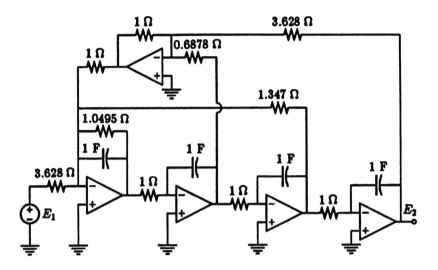

Figure 11.11: A fourth-order Chebyshev lowpass filter.

If $m = n$, a direct feed from the input to the output summer will supply the leading term in the numerator. To be specific, we shall assume that n is even.[6] To accomplish the realization we use the arrangement of Fig. 11.13 which yields

$$E_o = a_n E_1 - \frac{c_{n-1}s^{n-1} - c_{n-2}s^{n-2} + \cdots + c_1 s - c_0}{s^n + b_{n-1}s^{n-1} + b_{n-2}s^{n-2} + \cdots + b_1 s + b_0} E_1$$

or

$$\frac{E_o}{E_1} = [a_n s^n + (a_n b_{n-1} - c_{n-1})s^{n-1} + (a_n b_{n-2} + c_{n-2})s^{n-2}$$

$$+ \cdots + (a_n b_1 - c_1)s + (a_n b_0 + c_0)] \Big/$$

$$(s^n + b_{n-1}s^{n-1} + b_{n-2}s^{n-2} + \cdots + b_1 s + b_0) \qquad (11.17)$$

Equating the numerator coefficients of (11.17) to those in (11.1), we obtain

$$c_{n-1} = -a_{n-1} + a_n b_{n-1}$$

---

[6]The reader should have no difficulty modifying the sign pattern if n is odd instead.

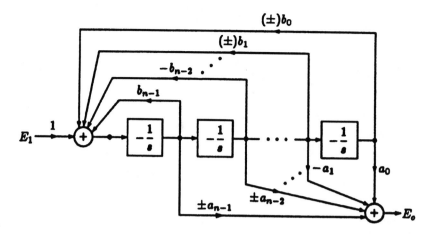

Figure 11.12: A scheme to realize transfer function (11.1) if $m < n$.

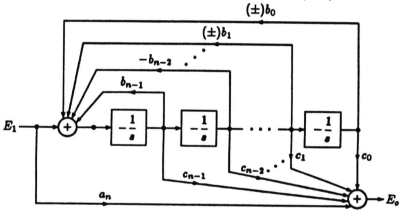

Figure 11.13: A scheme to realize transfer function (11.1) if $m = n$.

$$c_{n-2} = a_{n-2} - a_n b_{n-2}$$

$$\vdots \tag{11.18}$$

$$c_1 = -a_1 + a_n b_1$$

$$c_0 = a_0 - a_n b_0$$

EXAMPLE 11.6. Realize the lowpass elliptic filter of Example 11.4 using the block diagram of Fig. 11.13. Make the dc gain unity.

SOLUTION   Multiplying out the factors in (11.8) and scaling the numerator coefficients so that $a_0 = b_0$, we have

$$\frac{E_o}{E_1} = \frac{0.011860s^4 + 0.173572s^2 + 0.363848}{s^4 + 0.938796s^3 + 1.51553s^2 + 0.805178s + 0.363848} \qquad (11.19)$$

The denominator coefficients are implemented in a similar way as they are done in Example 11.5, Fig. 11.11. To obtain the forward gains, we apply (11.18) to get

$$c_3 = -a_3 + a_4 b_3 = 0.011134$$

$$c_2 = a_2 - a_4 b_2 = 0.15560$$

$$c_1 = -a_1 + a_4 b_1 = 0.0095495$$

$$c_0 = a_0 - a_4 b_0 = 0.35953$$

The filter circuit is shown in Fig. 11.14.

11.2.3   Realization using lossy integrators

The scheme for realizing an all-pole or general transfer function of any order can easily be modified if lossy integrators are used in place of the lossless ones. We shall illustrate the procedure by realizing a fourth-order all-pole transfer function. The pattern of this scheme should be apparent from this specific derivation. The reader should be able to extend this idea to transfer functions of any order.

The arrangement to realize a fourth-order all-pole function is shown in Fig. 11.15. We wish to realize

$$H(s) = \frac{E_2}{E_1} = \frac{a_0}{s^4 + b_3 s^3 + b_2 s^2 + b_1 s + b_0} \qquad (11.20)$$

For the summer, we have

$$\left(\frac{s+\beta}{\alpha}\right)^4 E_2 = f_4 E_1 - f_3 \left(\frac{s+\beta}{\alpha}\right)^3 E_2 + f_2 \left(\frac{s+\beta}{\alpha}\right)^2 E_2 - f_1 \left(\frac{s+\beta}{\alpha}\right) E_2 + f_0 E_2$$

Thus

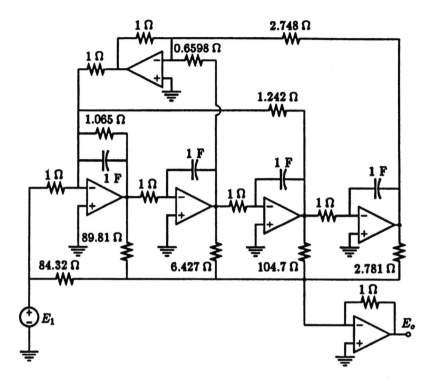

Figure 11.14: A circuit realizing the transfer function of (11.10).

$$\frac{E_2}{E_1} = \frac{f_4}{\left(\dfrac{s+\beta}{\alpha}\right)^4 + f_3 \left(\dfrac{s+\beta}{\alpha}\right)^3 - f_2 \left(\dfrac{s+\beta}{\alpha}\right)^2 + f_1 \left(\dfrac{s+\beta}{\alpha}\right) - f_0}$$

$$= (f_4\alpha^4) \Big/ \Big[s^4 + (4\beta + \alpha f_3)s^3 + (6\beta^2 + 3\alpha\beta f_3 - \alpha^2 f_2)s^2 + (4\beta^3$$

$$+ 3\alpha\beta^2 f_3 - 2\alpha^2\beta f_2 + \alpha^3 f_1)s + \beta^4 + \alpha\beta^3 f_3 - \alpha^2\beta^2 f_2 + \alpha^3\beta f_1 - \alpha^4 f_0 \Big]$$

To realize (11.20), we equate the corresponding coefficients. Thus

$$f_4\alpha^4 = a_0$$

$$4\beta + \alpha f_3 = b_3$$

$$6\beta^2 + 3\alpha\beta f_3 - \alpha^2 f_2 = b_2$$

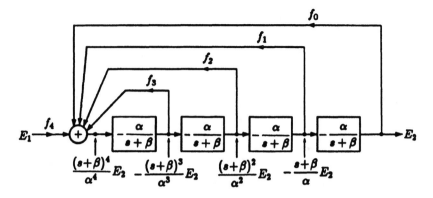

Figure 11.15: Realization of the fourth-order all-pole transfer function.

$$4\beta^3 + 3\alpha\beta^2 f_3 - 2\alpha^2\beta f_2 + \alpha^3 f_1 = b_1$$

$$\beta^4 + \alpha\beta^3 f_3 - \alpha^2\beta^2 f_2 + \alpha^3\beta f_1 - \alpha^4 f_0 = b_0$$

Now we have seven parameters and five equations to satisfy. Two of the parameters can be chosen arbitrarily. If we choose the values of α and β in advance, then the f's can be determined successively as follows:

$$f_4 = \frac{a_0}{\alpha^4}$$

$$f_3 = \frac{b_3 - 4\beta}{\alpha}$$

$$f_2 = \frac{-6\beta^2 + 3b_3\beta - b_2}{\alpha^2} \tag{11.21}$$

$$f_1 = \frac{-4\beta^3 + 3b_3\beta^2 - 2b_2\beta + b_1}{\alpha^3}$$

$$f_0 = \frac{-\beta^4 + b_3\beta^3 - b_2\beta^2 + b_1\beta - b_0}{\alpha^4}$$

EXAMPLE 11.7. Realize a normalized fourth-order Chebyshev low-pass filter with $\omega_0 = 1$ and $\alpha_p = 0.5$ dB, using lossy integrators with the transfer function $-1/(s+1)$.

SOLUTION From Table A.6 or using MATLAB command **cheb1ap**, we have

$$\frac{E_2}{E_1} = \frac{a_0}{s^4 + 1.19739s^3 + 1.71687s^2 + 1.02546s + 0.379051}$$

Since $\alpha = \beta = 1$, we obtain, using (11.21),

$$f_3 = -2.80262 \qquad f_2 = -4.12473$$

$$f_1 = -2.8162 \qquad f_0 = -0.87309$$

The op amp implementation of the block diagram of Fig. 11.15 with these coefficients is shown in Fig. 11.16.

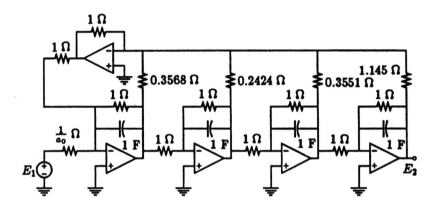

Figure 11.16: Chebyshev lowpass filter using lossy integrators.

It is seen that in Fig. 11.15 voltages $(-1)^i \left(\frac{s+\beta}{\alpha}\right)^i E_2$, $i = 1, 2, 3$, are all available. Any general fourth-order transfer function can be realized by techniques similar to those used in Figs 11.12 and 11.13 by properly summing these node voltages, and E_1 if $m = n$.

Also, inverting first-order lossy integrators are assumed in Fig. 11.15. If noninverting ones are used, the integrator output voltages will be $\left(\frac{s+\beta}{\alpha}\right)^i E_2$, $i = 1, 2, 3, 4$. If the signs of f_1 and f_3 are reversed, the overall transfer function will be unchanged. Hence noninverting lossy integrators may also used in this method of realization.

11.3   Lowpass-to-bandpass transformation

The lowpass-to-bandpass transformation described in Section 4.2 transforms a lowpass transfer function into a bandpass transfer function. As a matter of fact, we have already used it in Example 11.3. But the element replacement associated with this transformation, summarized in Table 4.1, will require that every capacitor be replaced by the parallel combination of a capacitor and an inductor. Hence the element replacement aspect of this transformation is not suitable for RC-op amp filters.

There are situations in which a lowpass active filter can be transformed into a bandpass filter by replacing certain functional blocks with new ones. We shall devote this section to two of these situations.

11.3.1   Coupled biquads with infinite Q

Lowpass filters obtained by the state-variable method of realizing a system function described in Section 11.2.1 can be transformed into bandpass filters. Take the block diagram of Figs 11.8 and 11.9. The only parts of the block diagrams that contain the variable s are the integrators. If we apply the lowpass-to-bandpass transform of (4.4)

$$s = \frac{S^2 + \Omega_0^2}{BS} \tag{11.22}$$

The integrator gain $\pm 1/s$ becomes

$$H_1(S) = \pm \frac{BS}{S^2 + \Omega_0^2} \tag{11.23}$$

which is the transfer function of a bandpass biquad with infinite Q. Hence if we replace every integrator of a lowpass filter with a bandpass biquad with infinite Q, the lowpass-to-bandpass transformation works just as well as in passive filters.

Biquads with infinite Q by themselves are unstable. They are actually oscillators. However, when they are used with the proper negative feedbacks, as we are doing in these configurations, they will function as biquads properly.

EXAMPLE 11.8. Design a normalized fourth-order Butterworth bandpass filter with $\Omega_0 = 1$, $B = 0.2$, and $G = 1$, by using a second-order lowpass prototype filter with noninverting lossless integrators and the lowpass-to-bandpass transformation.

SOLUTION   From Table A.2, we have

$$H_{\text{LP}}(s) = \frac{a_0}{s^2 + 1.41421s + 1} \tag{11.24}$$

By using the controller canonic form of Fig. 11.8, the block diagram that realizes the transfer function of (11.24) is shown in Fig. 11.17.

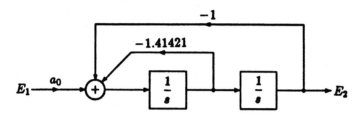

Figure 11.17: Realization of the transfer function of (11.24).

Applying the transformation

$$s = \frac{S^2 + 1}{0.2\,S}$$

to the filter, we should replace each integrator with a biquad with the transfer function

$$H_1(S) = \frac{0.2\,S}{S^2 + 1} \tag{11.25}$$

This bandpass biquad cannot be realized with either the Sallen-Key or the MFB second-order filters. We shall use the Tow-Thomas biquad. We choose $C_1 = C_2 = 1$ F and $R_2 = R_3$. From (10.61), we obtain

$$R_1 = \infty \qquad R_2 = R_3 = \frac{1}{\sqrt{b_0}} = 1\,\Omega \qquad R_4 = \frac{R_3}{G} = 5\,\Omega$$

The biquad is shown in Fig. 11.18. Hence the arrangement of Fig. 11.19 realizes the specified bandpass filter.[7]

---

[7]Since the input of the biquad is an integrator, the external inverting summer can be combined with the first integrator of the first biquad to save one op amp. If this is done, two feedback resistors - R_3 of the first biquad and the 0.7071-Ω feedback resistor - will be in parallel and can be combined. The overall gain will have the sign inverted. We leave the circuit as it is so the roles of each component will be clear.

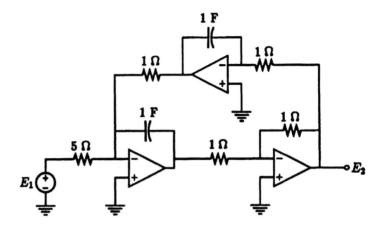

Figure 11.18: Biquad that realizes (11.25).

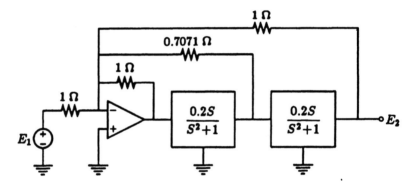

Figure 11.19: Two bandpass biquads coupled to form a fourth-order bandpass filter.

11.3.2 Bandpass filters using the primary resonator blocks

Our primary objective in Section 11.2.3, in which lossy integrators are used, is not to realize the all-pole transfer functions. Rather, it is the basis of another method of coupling a number of identical biquads with the same finite Q to realize high-order bandpass filters.

If we take the block diagram of Fig. 11.15 and apply the lowpass-to-bandpass transformation to each lossy integrator, its transfer function becomes

$$H_2(S) = -\frac{\alpha B\,S}{S^2 + \beta B\,S + \Omega_0^2} \tag{11.26}$$

Hence by using four identical bandpass biquads, we can realize an eighth-order bandpass filter. These identical biquads are known as the *primary resonator blocks* [Hu].

The reader should not have any difficulty extending this technique to bandpass filters of any order.

EXAMPLE 11.9. Transform the lowpass filter of Example 11.7 into a bandpass one with $\Omega_0 = 1$ and $B = 0.1$.

SOLUTION Since each lossy integrator has the transfer function

$$H(s) = -\frac{1}{s+1}$$

after applying the transform

$$s = \frac{S^2 + 1}{0.1\,S}$$

each integrator is replaced with a biquad with the transfer function

$$H_2(S) = -\frac{0.1\,S}{S^2 + 0.1\,S + 1}$$

The four biquads are coupled as shown in Fig. 11.20. Each block with the transfer function $H_2(S)$ can be realized by any suitable biquad described in Chapter 10, or from other sources. If noninverting biquads are used, the signs of every other feedback gain must alternate. Only one additional inverter would be required as described in connection with Fig. 11.10.[8]

11.4   Internal gain change

In the development of the state-variable type of technique of realizing high-order transfer functions, we have used block diagrams a great deal. This approach enables us to emphasize the functions of individual components and makes the mathematical derivation and visualization of the contribution of each component easier. These block diagrams are

---

[8]In this example, this inverter is not needed because all f's have the same sign.

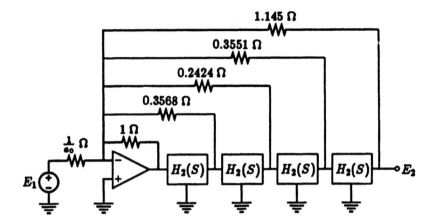

Figure 11.20: Arrangement of four primary resonator blocks to realize an eighth-order bandpass filter.

eventually implemented by using resistors, capacitors, op amps, op amp circuits, and biquads.

The block-diagram representation of a system has another advantage that is particularly useful in implementing op amp circuits. We can make internal changes without affecting the overall transfer function. This can be done by drawing an arbitrary closed surface that intersects a number of blocks and paths. If all gains directed into the surface are multiplied by a factor, and at the same time all gains directed away from the surface are multiplied by the reciprocal of this factor, the overall performance of the system remains unchanged. This is because the input-output relationship of the subsystem enclosed by the surface remains unchanged. This internal change can be made as many times as we wish. The factor can be either positive or negative. The purpose of these internal changes may be to change the time constants of certain integrators, to equalize the signal strengths throughout the system, or to reverse the signs of certain blocks or paths. These internal gain changes can be made more easily on a block diagram than on an op amp circuit.

Take the four-integrator system of Fig. 11.21(a). We could change the input gains by $-1/T_1$ and the output gains by $-T_1$ for all gains intersected by the closed surface (1). The result is the block diagram of Fig. 11.21(b). In Fig. 11.21(b), if we draw a closed surface (2) and simply make a sign change to all gains, the result is Fig. 11.21(c). In all three diagrams, E_o/E_1 remains unchanged. Note the last maneuver changes the sign of the output E_2.

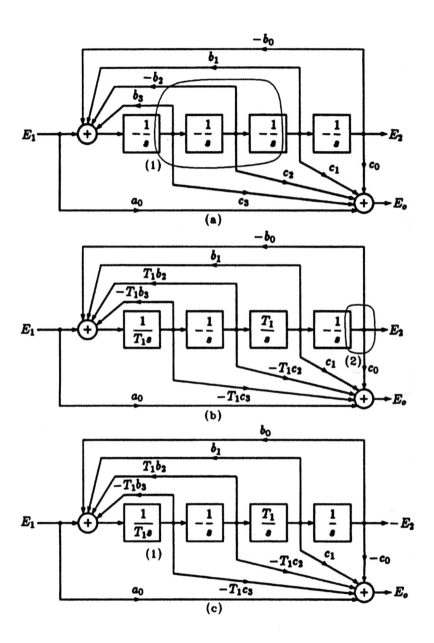

Figure 11.21: Examples of internal gain change.

11.5 Summary

In this chapter, we have developed two major groups of methods of realizing high-order active filters. One group is the cascade of lower-order sections - chiefly biquads. The other is the use of coupled integrators or biquads.

The cascade approach offers many advantages. The individual sections are easy to analyze, synthesize, and understand. These sections can each be designed, scaled, tuned, fabricated, and tested individually. Then there is the possibility that certain standard modules may be manufactured in advance with only a few key variable elements to be supplied externally to form many different filter sections. This can greatly reduce the cost of each section. Cascade realization is one of the most practical and popular methods of realizing active filters.

The major disadvantage of the cascade approach is the 'chain' nature of the arrangement. That is to say that any change or fluctuation in the performance of one section will directly affect the performance of the entire cascade. Thus the quality of the filter is limited by the 'weakest link' of the chain. The coupled-biquad method of realization includes feedbacks and, therefore, provides some reduction in the effect of each section on the overall performance of the filter. Hence the coupled-biquad approach tends to have lower sensitivity of the filter performance with respect to the variation of the individual biquads.

We include the state-variable methods of realizing high-order filters here as primarily of theoretical interest. These systems are usually quite sensitive when the order is high. They are also extremely difficult to tune as feedback gains are tied to certain coefficients rather than measurable parameters. But they do serve as a stepping stone to the coupled-biquad methods. Also, some of the concepts and techniques associated with this type of realization will be useful in the next chapter.

Problems

11.1 Design a fourth-order Butterworth lowpass filter with 3-dB frequency at 1 kHz. Use a cascade of two Sallen-Key biquads and use only 0.1 μF capacitors. The dc gain is arbitrary, but state the gain you have obtained. Use Design 1.

11.2 Repeat Problem 11.1 using Design 2.

11.3 Repeat Problem 11.1 using Design 3. If different, make the larger of the two capacitances 0.1 μF.

11.4 Repeat Problem 11.1 using Design 4. If different, make the larger of the two capacitances 0.1 μF.

11.5 Design a normalized Butterworth lowpass filter with $\alpha_p = 1$ dB, $\alpha_s \geq 20$ dB, and $\omega_s/\omega_p = 1.9$. The dc gain is arbitrary, but state the gain you have obtained. Normalize the pass band to 1 rad/sec. When biquads are called for, use Sallen-Key biquads with 1-F capacitors. Use Design 1.

11.6 Repeat Problem 11.5 using Design 2.

11.7 Repeat Problem 11.5 using Design 3. When Sallen-Key biquads are used, make the larger of the two capacitances 1 F.

11.8 Repeat Problem 11.5 using Design 4. When Sallen-Key biquads are used, make the larger of the two capacitances 1 F.

11.9 Design a fourth-order Chebyshev lowpass filter with $\alpha_p = 0.1$ dB, $\omega_p = 2\pi \times 500$ rad/sec, and the dc gain equal to unity. Use a cascade of two MFB biquads. Scale the impedances such that the larger of the two capacitances is 0.1 μF in each biquad and the capacitance value spread does not exceed 50:1.

11.10 Design a fourth-order Butterworth highpass filter with a high-frequency gain equal to 25 and 3-dB frequency at 500 Hz. Use a cascade of two Sallen-Key biquads with 0.02-μF capacitors.

11.11 Design a fourth-order Chebyshev highpass filter such that the high-frequency gain is 4, the passband ripple is 2 dB, and the pass band starts at 200 Hz. Use a cascade of two Sallen-Key highpass biquads with 0.1-μF capacitors and $\mu = 2$.

11.12 Design a fourth-order Butterworth bandpass filter such that the band center is at $2\pi \times 10^3$ rad/sec and the 3-dB bandwidth is $2\pi \times 100$ rad/sec. Use a cascade of two Sallen-Key (Design 1) bandpass biquads with 0.05-μF capacitors. Calculate the band-center gain you have obtained.

11.13 Repeat Problem 11.12 using one lowpass and one highpass biquad of Sallen-Key configuration.

11.14 Design a sixth-order Chebyshev bandpass filter with $\alpha_p = 0.5$ dB, $\Omega_0 = 2\pi \times 500$ rad/sec, and $B = 2\pi \times 100$ rad/sec. Use a cascade of three KHN biquads with 0.1-μF capacitors. Each biquad is used as a bandpass one. For each biquad, make $R_3 = R_5$. The gain is arbitrary. Calculate the gain at the center frequency you have obtained.

11.15 Repeat Problem 11.14 using one lowpass, one bandpass, and one highpass biquad.

11.16 Repeat Problem 11.14 using Tow-Thomas biquads. Make the gain at Ω_0 of each biquad equal to 3.

11.17 Design a normalized fourth-order elliptic lowpass filter with the maximum gain equal to unity, $\alpha_p = 1$ dB, $\omega_p = 1$ rad/sec, $\omega_s = 1.5$ rad/sec, using two Fleischer-Tow biquads with 1-F capacitors. Make the maximum gains of the two biquads equal.

11.18 Repeat the previous problem using KHN biquads.

11.19 Use four inverting integrators with time constant equal to 1 second to realize a fourth-order Chebyshev lowpass filter with $\omega_p = 1$ rad/sec, $\alpha_p = 0.5$ dB, and a maximum gain equal to unity. All capacitors are to be 1-F.

11.20 Repeat the previous problem using lossy integrators whose transfer functions are all $-1/(s+2)$.

11.21 Use four identical bandpass biquads as primary resonators to realize an eighth-order Butterworth bandpass filter with gain at the band center equal to 10, $\Omega_0 = 1$ rad/sec, and a 3-dB bandwidth of 0.1 rad/sec. Start with four lossy integrators with transfer functions equal to $-1/(s+\beta)$. Make $f_3 = 0$. Then apply an appropriate lowpass-to-bandpass transformation. It is sufficient to specify the voltage ratio of each resonator.

Chapter 12

Active simulation of passive filters

In addition to the methods presented in Chapter 11 for realizing high-order active filters, there is another entirely different approach to the same end. This other approach is to take a passive filter and simulate it by means of active devices and circuits. We shall devote this chapter to techniques in and examples of this approach.

There are at least two good reasons why this different approach can be attractive. One of them is the low sensitivity in the pass band of the doubly-terminated LC ladders, as established in Section 8.7. Hence we are starting out with a prototype network that has low sensitivities. Some of the reasons that lead to these low sensitivities can be expected to be carried over to the active simulated network.

The other reason is the wide availability of already compiled data and computer software on passive filters. There are many published tables that give the element values of singly-terminated and doubly-terminated LC ladder filters for many resistance ratios. Hence half of the work is already done for us and we need not duplicate the effort. All we have to do is pick out a suitable passive filter that satisfies our specification and simulate it with an active circuit.

Broadly speaking, there are two basically different methods of simulating an LC ladder filter. One of them is the *element replacement* method, in which we replace all inductors, either one by one or group by group, with active components. The other is first to study the internal relationships among the signals - voltages and currents. We then endeavor to arrange these internal relationships in such a way that each relationship can be effected by some op amp circuit. We shall call this method the *functional simulation* of LC ladder filters.

12.1   Some active twoports

As an introduction to some of the techniques described in this chapter, we shall describe several useful active twoports. This is done for historical interest as well as to put the device that we are going to use in its proper perspective. We shall use the notation for a twoport as shown in Fig. 12.1. Whenever the twoport of Fig. 12.1(b) is drawn, the port quantities in Fig. 12.1(a) are implied unless specifically indicated otherwise. For our purposes, a twoport is most conveniently characterized by

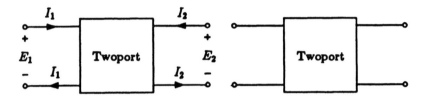

Figure 12.1: Notation used in a general twoport.

its *transmission, chain, or ABCD matrix*, $[F]$ in the equation

$$
\left[\begin{array}{c} E_1 \\ I_1 \end{array} \right] = \left[\begin{array}{cc} A & B \\ C & D \end{array} \right] \left[\begin{array}{c} E_2 \\ -I_2 \end{array} \right] = [F] \left[\begin{array}{c} E_2 \\ -I_2 \end{array} \right] \tag{12.1}
$$

The $[F]$ matrix of the four types of controlled sources should be familiar to the reader. We list them in Table 12.1 for easy reference.

Several active twoports are listed in Table 12.2. Some of these devices played important roles in the development of active filters. Some can still be used for active filter applications. Others are of only academic interest, although the understanding of their functions will help to pave the way for the material in the next section.

The names of the devices in Table 12.2 stem from their impedance transformation properties. Each name describes the relationship between the input impedance seen at one port and the terminating impedance at the other port. The arrangement in Fig. 12.2 has port 2 terminated in an impedance Z_L. Let us obtain the input impedance Z_{in} at port 1. We shall illustrate the impedance transformation property when the twoport is a positive impedance inverter. We have

$$
E_1 = \mp R_1 I_2 \qquad\qquad I_1 = \pm \frac{E_2}{R_2} \qquad\qquad E_2 = -Z_L I_2 \tag{12.2}
$$

Table 12.1: **The [F] matrices of controlled sources**

Controlled sources	$[F]$
Voltage-controlled current source	$\begin{matrix} 0 & -\dfrac{1}{g_m} \\ 0 & 0 \end{matrix}$
Current-controlled voltage source	$\begin{matrix} 0 & 0 \\ -\dfrac{1}{R_m} & 0 \end{matrix}$
Voltage-controlled voltage source	$\begin{matrix} \dfrac{1}{\mu} & 0 \\ 0 & 0 \end{matrix}$
Current-controlled current source	$\begin{matrix} 0 & 0 \\ 0 & \dfrac{1}{\alpha} \end{matrix}$

Table 12.2: **The [F] matrices of several active twoports**

Active twoport devices	$[F]$
Positive impedance inverter	$\begin{matrix} 0 & \pm R_1 \\ \pm\dfrac{1}{R_2} & 0 \end{matrix}$
Negative impedance inverter	$\begin{matrix} 0 & \pm R_1 \\ \mp\dfrac{1}{R_2} & 0 \end{matrix}$
Negative impedance converter	$\begin{matrix} \pm K_1 & 0 \\ 0 & \mp K_2 \end{matrix}$
Positive impedance converter	$\begin{matrix} \pm K_1 & 0 \\ 0 & \pm K_2 \end{matrix}$

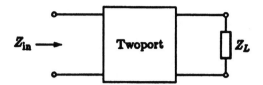

Figure 12.2: A terminated twoport.

Eliminating E_2 and I_2 from (12.2), we obtain

$$Z_{\text{in}} = \frac{E_1}{I_1} = \frac{R_1 R_2}{Z_L} \qquad (12.3)$$

Hence the input impedance is proportional to the reciprocal of the load impedance. Thus it is an impedance inverter. It's a *positive* inverter as the input impedance and the load impedance have the same sign. If $R_1 = R_2$, the positive impedance inverter is known as a *gyrator*.[1]

If the load is a capacitor of C farads, then the input impedance is

$$Z_{\text{in}} = R_1 R_2 C s \qquad (12.4)$$

which is the impedance of an inductor of $R_1 R_2 C$ henrys. This device can, therefore, be used to simulate an inductance. The realization of the devices listed in Table 12.2 is another major topic in itself [Su1]. We will not cover these topics here. Some of the circuits are included in the Problems section at the end of the chapter.

12.2   The generalized impedance converter (GIC)

The elements in the $[F]$ matrices of the active devices listed in Table 12.2 are all constants. These elements can be generalized to be functions of s. When this is done, the devices will be called *generalized* active devices. One of the most useful generalized devices is the generalized (positive) impedance converter (GIC). We shall write the $[F]$ matrix of a GIC as

$$[F] = \begin{bmatrix} f_1(s) & 0 \\ 0 & f_2(s) \end{bmatrix} \qquad (12.5)$$

---

[1] Actually, a gyrator is a lossless device. In microwave technology, this device is indeed a passive one. However, in electronic circuits, this device is usually realized using active circuit elements.

Among the many GIC circuits, the one that has proven to be most practical, versatile, and easy to implement, is the Antoniou GIC [An] shown in Fig. 12.3.

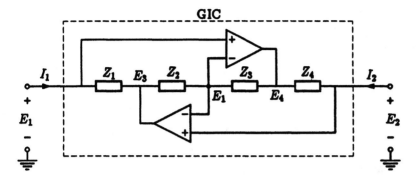

Figure 12.3: A generalized impedance converter.

In Fig. 12.3, because of the virtual short of the input terminals of the op amps, we readily see that

$$E_1 = E_2 \tag{12.6}$$

With the three interior node voltages as labeled, we can write three current equations:

$$I_1 = \frac{E_1 - E_3}{Z_1}$$

$$I_2 = \frac{E_2 - E_4}{Z_4} = \frac{E_1 - E_4}{Z_4}$$

$$\frac{E_3 - E_1}{Z_2} + \frac{E_4 - E_1}{Z_3} = 0$$

Solving, we obtain

$$I_1 = -\frac{Z_2 Z_4}{Z_1 Z_3} I_2 = -f_2(s) I_2 \tag{12.7}$$

Hence the circuit of Fig. 12.3 is a GIC with $f_1(s) = 1$ and

$$f_2(s) = f(s) = \frac{Z_2 Z_4}{Z_1 Z_3} \tag{12.8}$$

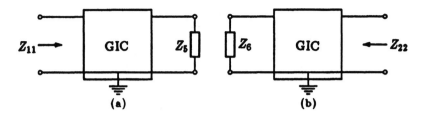

(a)   (b)

Figure 12.4: Terminated GIC's.

If we terminated one port in an impedance as shown in Fig. 12.4, the
input impedances seen looking into the other ports are

$$Z_{11} = \frac{E_1}{I_1} = \frac{1}{f(s)} \times \frac{E_2}{-I_2} = \frac{1}{f(s)} Z_5 \qquad (12.9)$$

$$Z_{22} = \frac{E_2}{I_2} = f(s) \times \frac{E_1}{-I_1} = f(s) Z_6 \qquad (12.10)$$

The impedance conversion properties of this device are now apparent.
Henceforth, we shall represent a GIC by its *impedance* conversion ratio
as shown in Fig. 12.5. The ratio indicated in each GIC is the *impedance*
ratio, somewhat similar to the turns ratio associated with an ideal transformer. It should be emphasized that (12.6) and (12.7) are implied in
these symbols.

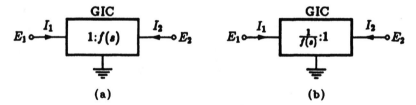

(a)   (b)

Figure 12.5: Simplified symbols for a GIC.

12.3   Simulation of inductances in an LC ladder

In this section, we shall show how we can take an LC ladder and replace all its inductances with GIC's and resistors. We shall start with
a very simple replacement method. Then we'll gradually generalize the
technique to apply to any general network.

12.3.1   Simulation of grounded inductors

If we terminate port 2 of a GIC with an impedance ratio $Ks{:}1$, where K is real, in a resistance R_L as shown in Fig. 12.6, the input impedance of port 1 will be

$$Z_{in} = KsR_L \tag{12.11}$$

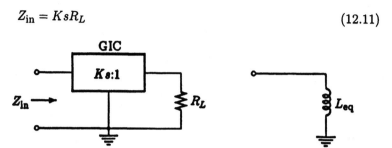

Figure 12.6: Simulation of a grounded inductor.

This input impedance is the same as that of a grounded inductor whose inductance is

$$L_{eq} = KR_L \tag{12.12}$$

To realize such a GIC, we could use the circuit of Fig. 12.3 and let either Z_2 or Z_4 be a capacitor and the other three impedances be resistors. (In the following formulas, the subscripts correspond to those of the Z's in Fig. 12.3.) Explicitly, from (12.8) and (12.9), we will have

$$Z_{in} = \frac{R_1 R_3 s C_2}{R_4} \times R_L \quad \text{or} \quad \frac{R_1 R_3 s C_4}{R_2} \times R_L \tag{12.13}$$

or

$$L_{eq} = \frac{R_1 R_3 C_2 R_L}{R_4} \quad \text{or} \quad \frac{R_1 R_3 C_4 R_L}{R_2} \tag{12.14}$$

which implies that

$$K = \frac{R_1 R_3 C_2}{R_4} \quad \text{or} \quad \frac{R_1 R_3 C_4}{R_2}$$

As an application of this simulation, whenever we need a grounded inductor, we can replace it with a circuit containing four resistors, two op amps, and one capacitor. For example, the circuit of Fig. 12.7 is a fifth-order Butterworth highpass filter. The active circuit of Fig. 12.8 performs the same filtering function as that in Fig. 12.7 if we assume the op amps to be ideal.

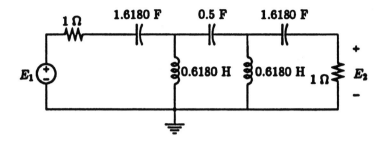

Figure 12.7: A fifth-order Butterworth highpass filter.

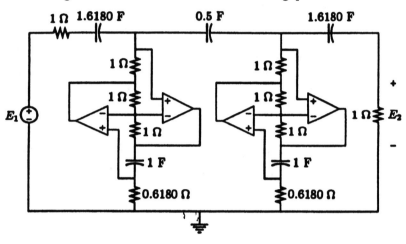

Figure 12.8: Active simulation of the filter of Fig. 12.7.

12.3.2   Simulation of floating inductors

Because of the ground implied in each op amp, the terminated GIC method of simulating an inductor described in the previous subsection renders an inductor that is inherently grounded at one of its terminals. The arrangement shown in Fig. 12.9 can be used to simulate an un-grounded inductor. In Fig. 12.9, from (12.6) and (12.7), we have

$$E_1' = E_1 \qquad E_2' = E_2$$

$$I_1' = KsI_1 \qquad I_2' = KsI_2$$

$$I_1' = -I_2' \qquad I_1 = -I_2$$

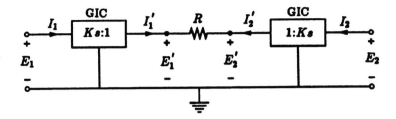

Figure 12.9: Simulation of a floating inductor.

Since

$$\frac{E_1' - E_2'}{I_1'} = R$$

It follows that

$$\frac{E_1 - E_2}{I_1} = KsR \tag{12.15}$$

In Fig. 12.10, we have

$$\frac{E_1 - E_2}{I_1} = sL_{eq} \qquad \text{and} \qquad I_1 = -I_2 \tag{12.16}$$

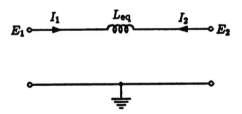

Figure 12.10: A grounded two-port consisting of a ungrounded inductor.

Hence the two twoports in Figs 12.9 and 12.10 are equivalent and

$$L_{eq} = KR \tag{12.17}$$

Although the ground is still present in Fig. 12.10, the relationships among the four quantities - I_1, I_2, E_1, and E_2 - associated with the top two terminals of the twoport are unaffected. In other words, a floating inductor has been simulated by this arrangement.

As an example of this method of realizing floating inductors, the lowpass filter of Fig. 12.11 can be simulated by the arrangement of Fig. 12.12 if we make

$$R_2 = \frac{L_2}{K} \quad \text{and} \quad R_4 = \frac{L_4}{K}$$

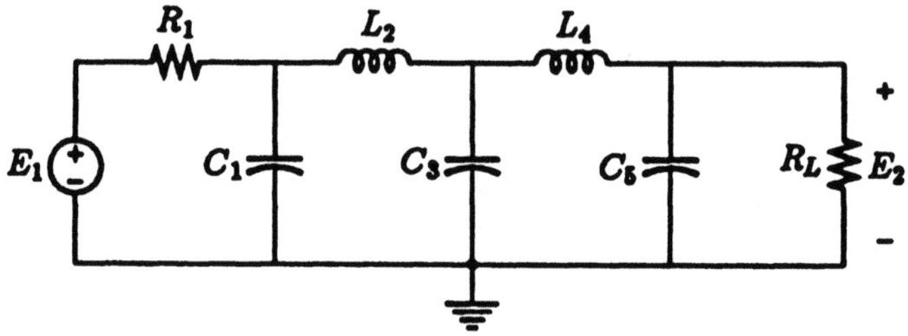

Figure 12.11: A lowpass filter.

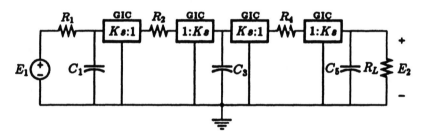

Figure 12.12: Active simulation of the filter of Fig. 12.11.

12.3.3   Simulation of groups of inductors

The method of simulating a floating inductor can be extended to simulating an inductor subnetwork. If we have a subnetwork that contains only a number of inductors, they can be replaced with a resistive subnetwork and a number of GIC's [Go]. Suppose the inductor subnetwork is connected to the rest of the network through n terminals as shown in Fig. 12.13(a). If subnetwork N is characterized by its impedance matrix $[Z]$, then we have

$$\begin{bmatrix} E_1 \\ E_2 \\ \vdots \\ E_n \end{bmatrix} = [Z] \begin{bmatrix} I_1 \\ I_2 \\ \vdots \\ I_n \end{bmatrix} \tag{12.18}$$

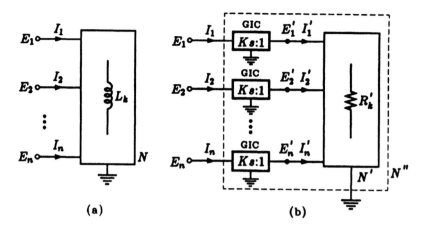

Figure 12.13: Simulation of a group of inductors.

In the arrangement of Fig. 12.13(b), we have

$$E'_k = E_k \qquad \text{and} \qquad I'_k = K s I_k$$

for $k = 1, 2, \ldots, n$. If network N' has an impedance matrix $[Z']$, then

$$
\begin{bmatrix} E'_1 \\ E'_2 \\ \vdots \\ E'_n \end{bmatrix} = [Z'] \begin{bmatrix} I'_1 \\ I'_2 \\ \vdots \\ I'_n \end{bmatrix}
\tag{12.19}
$$

and the impedance matrix of N'' in Fig. 12.13(b) will be $Ks[Z']$. Hence, if we make every element of $[Z']$

$$z'_{ij} = \frac{z_{ij}}{K s} \tag{12.20}$$

where z_{ij} is the ij element of $[Z]$, then the impedance matrix of N'' and that of N will be identical. The requirement given in (12.20) can be satisfied by replacing every inductance of L_k henrys in N with a resistance of R'_k ohms such that

$$R'_k = \frac{L_k}{K} \tag{12.21}$$

for all k. Then N'' and N are equivalent to each other. As an example of this method of simulation, the two inductors, L_2 and L_4, in Fig. 12.11

form a subnetwork that is connected to the rest of the filter through three terminals as shown in Fig. 12.14(a). Hence this subnetwork can be realized using three GIC's and two resistors as shown in Fig. 12.14(b), in which we also make $R_2 = L_2/K$ and $R_4 = L_4/K$. Thus the number of GIC's utilized is reduced by one as compared with the realization of Fig. 12.12.

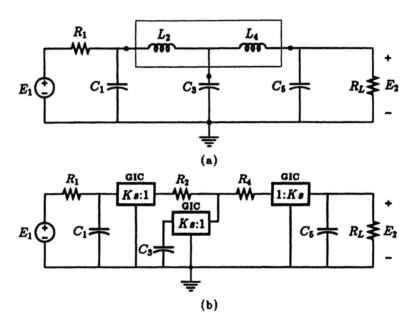

Figure 12.14: An alternative active simulation of the filter of Fig. 12.11.

As another example of this technique, the filter of Fig. 12.15 is realized by the RC-op amp network of Fig. 12.3.3. In Fig. 12.16, we need to make

$$R_2 = \frac{L_2}{K_1} \quad \text{and} \quad R_k = \frac{L_k}{K_2} \tag{12.22}$$

for $k = 3, 4, \ldots, 10$. The values of K_1 and K_2 may be the same or different.

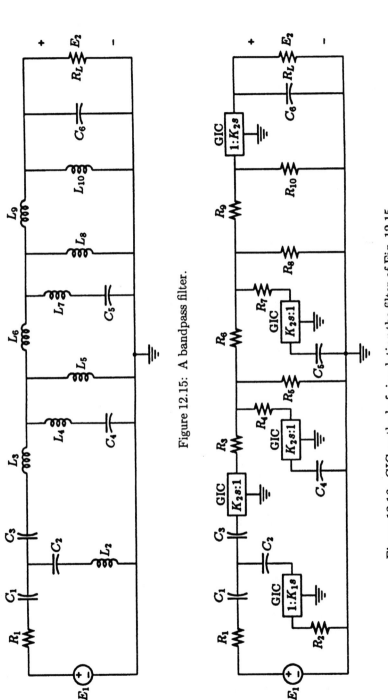

Figure 12.15: A bandpass filter.

Figure 12.16: GIC method of simulating the filter of Fig. 12.15.

12.4   Simulation using frequency-dependent negative resistances (FDNR's)

In the GIC circuit of Fig. 12.3, if we let both Z_1 and Z_3 be capacitors and Z_2 and Z_4 be resistors, then according to (12.8), for the GIC

$$f(s) = R_2 R_4 C_1 C_3 s^2 = \frac{s^2}{K} \tag{12.23}$$

If we terminate port 2 of such a GIC in a resistance R_L as shown in Fig. 12.17(a), the input impedance Z_{in} is

$$Z_{\text{in}}(s) = \frac{1}{f(s)} \cdot R_L = \frac{K}{s^2} \cdot R_L = \frac{R_L}{R_2 R_4 C_1 C_3} \cdot \frac{1}{s^2} = \frac{1}{Ds^2} \tag{12.24}$$

where

$$D = \frac{R_2 R_4 C_1 C_3}{R_L}$$

and has the dimension of $\Omega \times F^2$. Along the j axis,

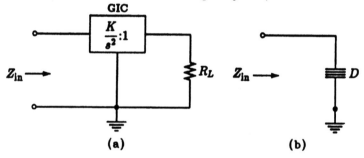

Figure 12.17: A grounded FDNR.

$$Z_{\text{in}}(j\omega) = -\frac{1}{D\omega^2} \tag{12.25}$$

This impedance is real and negative, and its value varies with frequency. This simulated impedance is known as a *frequency-dependent negative resistance* (FDNR) [Br]. An FDNR is represented by the symbol shown in Fig. 12.17(b).

In Section 1.6, it was shown that any voltage ratio E_2/E_1 is unchanged if all impedances in the network are scaled by the same factor k_z. In fact, this assertion will still be true if k_z is a function of s instead of a constant. In particular, if we let $k_z = 1/s$, then

(1) the impedance of a resistance of R ohms becomes R/s ohms, which is the impedance of a capacitor whose capacitance is $1/R$ farads,

(2) the impedance of an inductance of L henrys is sL ohms and will become L ohms, which is the impedance of a resistor whose resistance is L ohms, and

(3) the impedance of a capacitance of C farads is $1/sC$ ohms and will become $1/s^2C$ ohms, which is the impedance of an FDNR whose value for D is C ΩF^2.

As an example, we start with the highpass filter of Fig. 12.18 in which the impedance of each element is indicated. After impedance-scaling with $k_z = 1/s$, the network of Fig. 12.19 results. The two grounded FDNR's can be realized by the arrangement of Fig. 12.17. Fig. 12.20 shows the same circuit with the structure of the FDNR's explicitly shown.

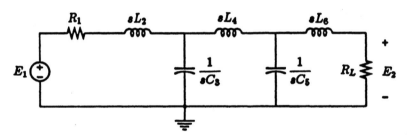

Figure 12.18: A lowpass filter. Element values are their impedances.

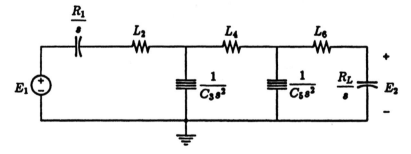

Figure 12.19: Filter of Fig. 12.18 after impedance-scaling by $1/s$. Element values are their impedances.

Analogous to realizing floating inductors and inductor subnetworks as described in the previous section, floating FDNR's and FDNR subnetworks can also be simulated by a number of GIC's and ungrounded

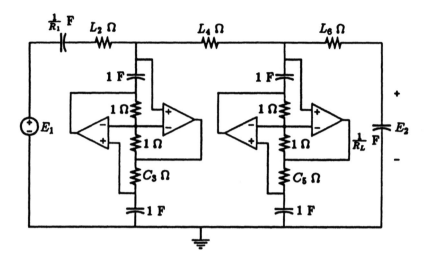

Figure 12.20: Circuit of Fig. 12.19 with details of FDNR's shown.

resistors and resistor subnetworks respectively.

12.5   Functional simulation of passive filters

In the previous sections, the simulation was accomplished by taking a passive filter and simulating the elements either individually or in groups. A different approach is to take a passive filter and write out the relationships that must be satisfied by the voltages and currents within the circuit. If this is done in an appropriate manner such that these relationships can be reproduced by an active circuit, the active circuit will duplicate the performance of the passive filter.

12.5.1   The leap-frog realization

Take the ladder network of Fig. 12.21, in which the following relationships must be satisfied.

$$I_1 = Y_1(E_1 - E_2) \qquad E_2 = Z_2(I_1 - I_3)$$

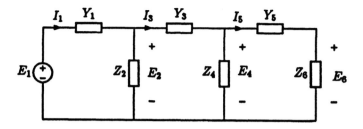

Figure 12.21: A ladder network.

$$I_3 = Y_3(E_2 - E_4) \qquad\qquad E_4 = Z_4(I_3 - I_5) \qquad\qquad (12.26)$$

$$I_5 = Y_5(E_4 - E_6) \qquad\qquad E_6 = Z_6 I_5$$

Suppose the relevant relationship of this ladder network is the voltage ratio E_6/E_1. All quantities in (12.26) other than E_1 and E_6 do not have to be tied to their original identities. For example, all I's may be replaced with another set of voltages and all Y's and Z's may be regarded as functions of s that relate the voltages in the equations. Then the relationship between E_6 and E_1 will remain unchanged. In other words, we rewrite (12.26) as

$$V_1 = f_1(s)(E_1 - E_2) \qquad\qquad E_2 = f_2(s)(V_1 - V_3)$$

$$V_3 = f_3(s)(E_2 - E_4) \qquad\qquad E_4 = f_4(s)(V_3 - V_5) \qquad (12.27)$$

$$V_5 = f_5(s)(E_4 - E_6) \qquad\qquad E_6 = f_6(s)V_5$$

where $f_1(s) = Y_1$, $f_2(s) = Z_2$, etc. Equation (12.27) can be regarded as a set of relationships that must be satisfied by a system in order to obtain the same voltage transfer function E_6/E_1 for the LC ladder of Fig. 12.21. If we can construct a system that enforces the relationships in (12.27), this system will duplicate the function of the ladder of Fig. 12.21.

The relationships in the block diagram of Fig. 12.22(a) satisfy all these relationships. Therefore, the relationship between E_6 and E_1 in Figs 12.21 and 12.22(a) are identical. The pattern of the block diagram of Fig. 12.22(a) is redrawn in Fig. 12.22(b). The latter shows the resemblance of this arrangement to a children's game. This realization is commonly known as the *leap-frog realization* [GG]. The success of this method depends not only on our ability to write the internal relationships in such a way that they are amenable to the step-by-step representation

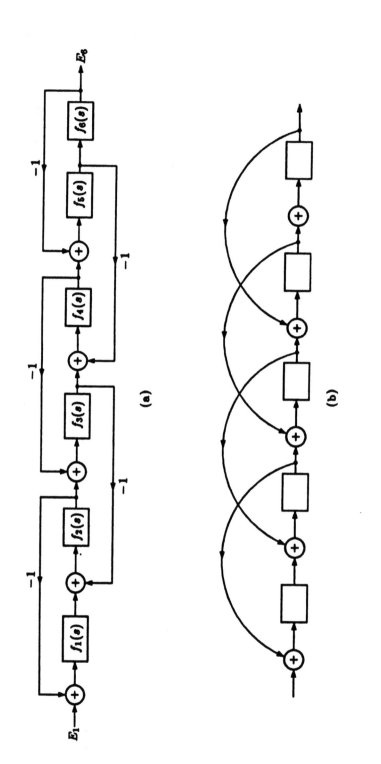

Figure 12.22: Block diagram of the functional simulation of the ladder network of Fig. 12.21.

of each equation but also on the practical realizability of the functions $f_k(s)$ in those equations as transfer functions.

The arrangement of Fig. 12.5.1(a) suffers two drawbacks from the point of view of realizing the block diagram with RC-op amp circuits. One is the fact that an inverter is required in each of the feedback paths. The other is that all integrators are noninverting. Both of these shortcomings can be alleviated by making an internal change to the block diagram. The general principle of internal gain change was explained in Section 11.4. We now consider the two closed surfaces shown in Fig. 12.23. If all gains going into each closed surface are multiplied by $\alpha = -1$ and, at the same time, all gains leaving the same closed surface are multiplied by $1/\alpha = -1$, the overall performance of the system is obviously unchanged (except for a sign reversal). The result is the block diagram shown in Fig. 12.24. The latter can be implemented using fewer inverters.

12.5.2 Leap-frog realization of the lowpass LC ladder

In this section, we shall show, by way of an example, how the leap-frog method of realization can be applied to a lowpass LC ladder. We shall take the passive filter of Fig. 12.18, and then write

$$f_1(s) = \frac{1}{R_1 + sL_2} = \frac{1/L_2}{s + R_1/L_2} \qquad f_2(s) = \frac{1}{sC_3}$$

$$f_3(s) = \frac{1}{sL_4} \qquad f_4(s) = \frac{1}{sC_5} \qquad (12.28)$$

$$f_5(s) = \frac{1}{sL_6} \qquad f_6(s) = R_L$$

Now if each block in Fig. 12.24 is implemented according to the transfer functions given in (12.28), with negative signs included as needed, the transfer function of Fig. 12.18 is duplicated. This will require that each block be either a lossy or lossless integrator, some of them inverting and others noninverting. The final op amp implementation of the block diagram of Fig. 12.24 is given in Fig. 12.25, in which most resistance values are normalized to 1 Ω.

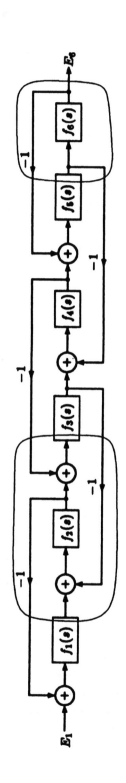

Figure 12.23: Closed surfaces for internal sign changes.

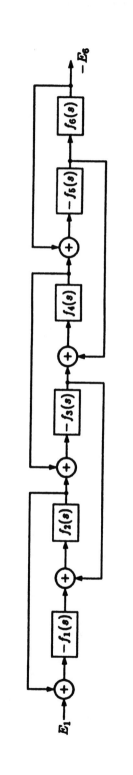

Figure 12.24: An alternative block diagram that simulates the filter of Fig. 12.21.

12.5.3 Leap-frog realization of bandpass filters using bi-quads

As was shown in Section 4.2, a bandpass filter can be obtained from a lowpass prototype by replacing every s with

$$s = \frac{S^2 + \omega_0^2}{BS} \tag{12.29}$$

This transformation can be effected by applying the transformation to the individual integrator blocks in the leap-frog simulation of the lowpass prototype. We shall look at the example of the simulation of the lowpass filter of Fig. 12.18 with the block diagram of Fig. 12.5.2. The transfer functions are given in (12.28). The block that contains $f_1(s)$ should now be replaced with

$$f_1(s) = \frac{1}{R_1 + \left(\dfrac{S^2 + \omega_0^2}{BS}\right)L_2} = \frac{\dfrac{B}{L_2}S}{S^2 + \dfrac{BR_1}{L_2}S + \omega_0^2} \tag{12.30}$$

which is the transfer function of a bandpass biquad. The other four integrator blocks all become

$$f_k(s) = \frac{1}{\left(\dfrac{S^2 + \omega_0^2}{BS}\right)x} = \frac{\dfrac{B}{x}S}{S^2 + \omega_0^2} \tag{12.31}$$

for which $x = C_3$ if $k = 2$, $x = L_4$ if $k = 3$, $x = C_5$ if $k = 4$, and $x = L_6$ if $k = 5$. These blocks are each a bandpass biquad with infinite pole-Q.

Hence, the simulations of lowpass filters using lossy and lossless integrators can be transformed into bandpass filters by replacing each integrator with a suitable bandpass biquad. Each biquad can be realized by using any of the suitable circuits in Chapter 10, or from any other source.

12.5.4 Leap-frog realization of bandpass filters using integrators

As an alternative, we could also first obtain the bandpass LC ladder by using the element replacements summarized in Table 4.1. Figure 12.26 is such a circuit obtained from a third-order lowpass prototype. The internal relationships are

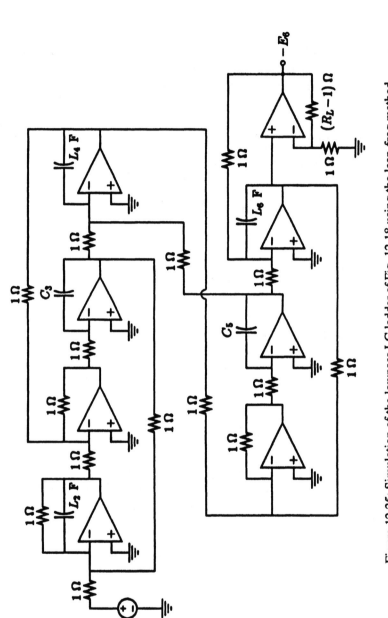

Figure 12.25: Simulation of the lowpass LC ladder of Fig. 12.18 using the leap-frog method.

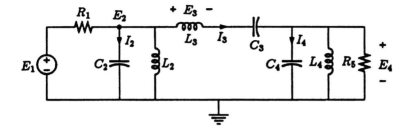

Figure 12.26: A sixth-order bandpass filter.

$$I_2 = \frac{E_1 - E_2}{R_1} - \frac{E_2}{sL_2} - I_3 \qquad E_2 = \frac{I_2}{sC_2}$$

$$E_3 = E_2 - E_4 - \frac{I_3}{sC_3} \qquad I_3 = \frac{E_3}{sL_3} \qquad (12.32)$$

$$E_4 = \frac{I_4}{sC_4} \qquad I_4 = I_3 - \frac{E_4}{sL_4} - \frac{E_4}{R_5}$$

Now we construct a block diagram to realize the equations given in (12.32). The block diagram is shown in Fig. 12.27.

In order to reduce the number of inverters required, we perform an internal sign change. Two surfaces were drawn in Fig. 12.27.

Reversing the signs of all gains that cross these surfaces, we obtain the block diagram shown in Fig. 12.28. We now implement this block diagram with R, C, and op amps. The result is the circuit shown in Fig. 12.29.

12.5.5 Simulation of a special bandpass filter

A slight variation of the method shown in this section can sometimes be effected on certain specially designed bandpass filters. (We are referring to certain bandpass filters that are not obtained by using the lowpass-to-bandpass transformation. They are designed directly using special techniques.) An example is shown in Fig. 12.30 [Sz].

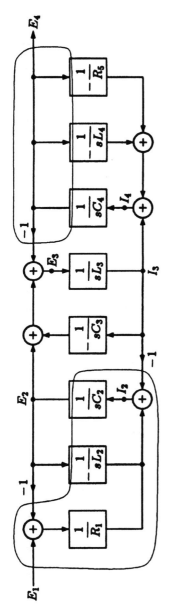

Figure 12.27: Block diagram realizing the equations in (12.32).

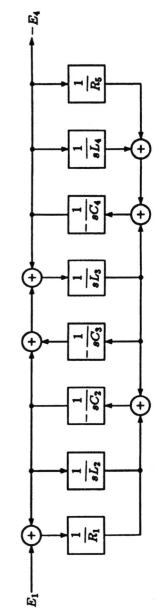

Figure 12.28: Block diagram of Fig. 12.27 with some internal sign changes.

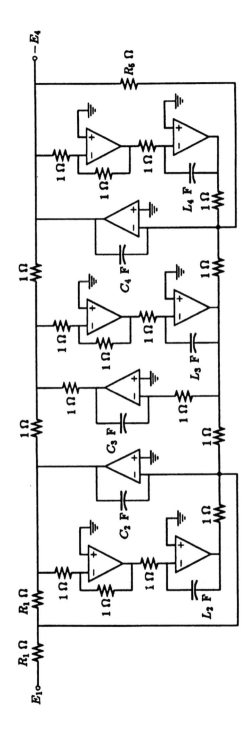

Figure 12.29: Op amp circuit implementation of the block diagram of Fig. 12.28.

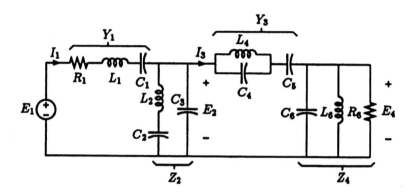

Figure 12.30: A special passive bandpass filter.

This filter is a two-rung ladder - the ladder network of Fig. 12.21 with
Y_5 and Z_6 removed - and the output is taken as E_4. The following
relationships are still true as in (12.26):

$$I_1 = Y_1(E_1 - E_2)$$

$$E_2 = Z_2(I_1 - I_3)$$

$$I_3 = Y_3(E_2 - E_4) \tag{12.33}$$

$$E_4 = Z_4 I_3$$

But we have

$$Y_1 = \frac{1}{L_1} \times \frac{s}{s^2 + \dfrac{R_1}{L_1}s + \dfrac{1}{L_1 C_1}}$$

$$Z_2 = \frac{1}{C_3} \times \frac{s^2 + \dfrac{1}{L_2 C_2}}{s\left(s^2 + \dfrac{1}{L_2 C_2} + \dfrac{1}{L_2 C_3}\right)}$$

$$Y_3 = \frac{L_4 C_4 C_5}{L_4(C_4 + C_5)} \times \frac{s\left(s^2 + \dfrac{1}{L_4 C_4}\right)}{s^2 + \dfrac{1}{L_4(C_4 + C_5)}} \tag{12.34}$$

$$Z_4 = \frac{1}{C_6} \times \frac{s}{s^2 + \dfrac{1}{R_6 C_6}s + \dfrac{1}{L_6 C_6}}$$

If we identify Y_1, Z_2, Y_3, and Z_4 as the transfer functions as before, third-order filter sections will be required. To avoid using third-order filter sections, we can rewrite (12.33) as

$$\frac{I_1}{s} = \frac{Y_1}{s}(E_1 - E_2)$$

$$E_2 = sZ_2\left(\frac{I_1}{s} - \frac{I_3}{s}\right)$$

$$\frac{I_3}{s} = \frac{Y_3}{s}(E_2 - E_4) \tag{12.35}$$

$$E_4 = sZ_4\left(\frac{I_3}{s}\right)$$

Then we can let

$$V_1 = \frac{I_1}{s} \quad \text{and} \quad V_2 = \frac{I_2}{s} \quad \text{and} \quad V_3 = \frac{I_3}{s} \tag{12.36}$$

and

$$f_1(s) = \frac{Y_1}{s} = \frac{1}{L_1} \times \frac{1}{s^2 + \dfrac{R_1}{L_1}s + \dfrac{1}{L_1C_1}}$$

$$f_2(s) = Z_2 = \frac{1}{C_3} \times \frac{s^2 + \dfrac{1}{L_2C_2}}{s^2 + \dfrac{1}{L_2C_2} + \dfrac{1}{L_2C_3}}$$

$$f_3(s) = \frac{Y_3}{s} = \frac{L_4C_4C_5}{L_4(C_4 + C_5)} \times \frac{s^2 + \dfrac{1}{L_4C_4}}{s^2 + \dfrac{1}{L_4(C_4 + C_5)}} \tag{12.37}$$

$$f_4(s) = Z_4 = \frac{1}{C_6} \times \frac{s^2}{s^2 + \dfrac{1}{R_6C_6}s + \dfrac{1}{L_6C_6}}$$

Then the filter of Fig. 12.30 can be simulated by the system shown in Fig. 12.31 in which each rectangular block can be realized by a biquad with its transfer function specified in (12.37).

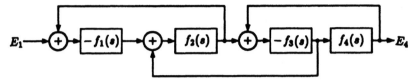

Figure 12.31: Active simulation of the filter of Fig. 12.30.

12.6 Summary

In this chapter we have described several methods for realizing active filters. These methods are all based on the simulation of known passive filters using active network techniques. Two major approaches are included. One of them is to replace impractical passive elements, chiefly inductances, with active circuits. The other is based on the internal electrical relationships of passive filters. When these relationships are written in such a way that they can be implemented with active circuits, the functional simulation of these passive filters will have been achieved. Experience has proven that these techniques can also provide practical and attractive alternatives to those methods described in the previous chapter.

The reader should keep in mind that the various methods represented in the last two chapters merely represent simpler and the more popular approaches of generating active filters. Numerous other methods of synthesizing active filters exist and some of them can also prove to be very practical and useful. However, we cannot include all of them in the limited space and time we are able to devote to these subjects.

Problems

12.1 For the following circuit, express the input impedance Z_{in} in terms of Z_1 and Z_2.

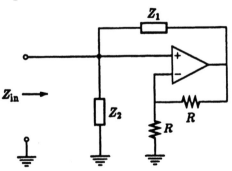

12.2 Obtain the chain matrix of the following twoport. What is this twoport as an impedance transformation device?

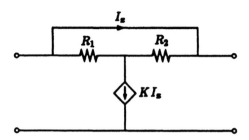

12.3 Obtain the chain matrix of the following twoport. What is this twoport as an impedance transformation device?

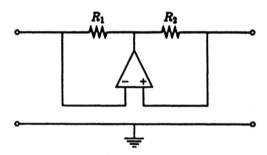

12.4 The following impedance function is not a lossless function. However, it can be decomposed into the difference of two lossless functions by first obtaining Foster's expansion of the function. Use the device of the previous problem with $R_1 = 1 \ \Omega$ and $R_2 = 2 \ \Omega$ to realize this impedance.

$$Z(s) = \frac{(s^2 + 2)(s^2 + 4)}{s(s^2 + 1)(s^2 + 5)} \ \Omega$$

12.5 (a) What active device is the following twoport? (b) If an impedance Z_L is connected between the two terminals of port 2, what would be the input impedance at port 1?

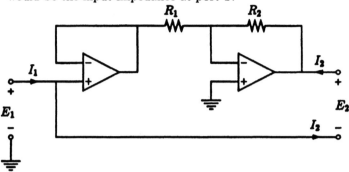

12.6 The following circuit is known as the Riordan impedance converter [Ri]. Obtain the input impedance Z_{in} in terms of the Z's.

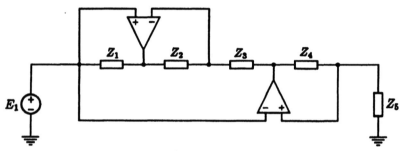

12.7 Twoport N is a positive impedance inverter whose chain matrix is given in Table 12.2. Obtain E_2/E_1.

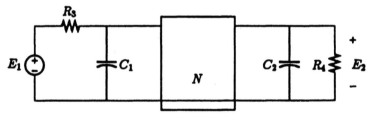

12.8 Realize a grounded impedance equal to $50s^2$ Ω. Give the schematic diagram of your circuit.

12.9 The voltage ratio of the network in (a) is

$$\frac{E_2}{E_1} = \frac{\dfrac{1}{RC}s}{s^2 + \dfrac{1}{RC}s + \dfrac{1}{LC}}$$

which is the transfer function of a bandpass biquad with $G = 1$. In (b), the inductance is simulated by a GIC. This circuit has the shortcoming that the output impedance can be quite high. It would be much better to take the output at E_o. Show that the gain E_o/E_1 is a constant times E_2/E_1. What is this constant?

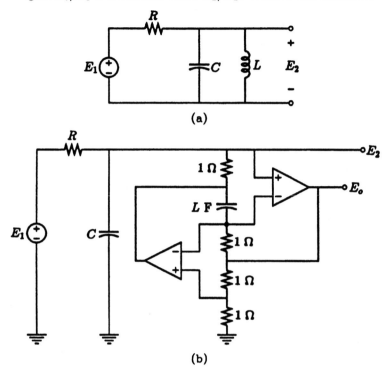

(a)

(b)

12.10 Obtain the transfer function E_o/E_1 of the following circuit.

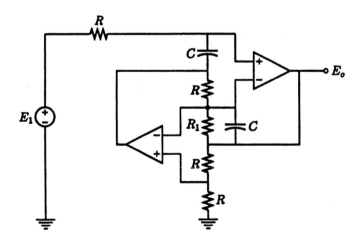

12.11 Obtain E_2/E_1 and E_o/E_1 for the circuit (b) in Problem 10.9 if R and C are interchanged.

12.12 Simulate the inductors in the following circuit with resistors and GIC's. Use as few GIC's as possible. For each GIC, it is sufficient to indicate its impedance ratio.

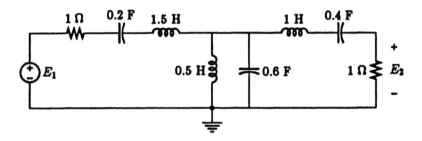

12.13 It is desired to replace all the capacitors in the following circuit using GIC's and resistors. Use as few GIC's as possible. For each GIC, it is sufficient to indicate its impedance ratio.

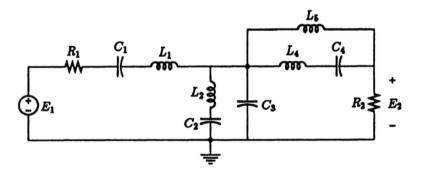

12.14 It is desired to simulate the following passive filter using only resistors, capacitors, and GIC's with an impedance ratio of $s{:}1$. Do this using as few GIC's as possible.

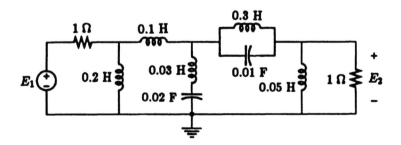

12.15 A passive filter has already been designed and is represented in the first diagram below. It is desired to insert three GIC's at points A, B, and C and replace some of the elements with different elements as shown in the second diagram. The impedance ratios of the GIC's are given as shown. Specify all impedances in the second diagram so that the voltage ratio E_2/E_1 will remain unchanged.

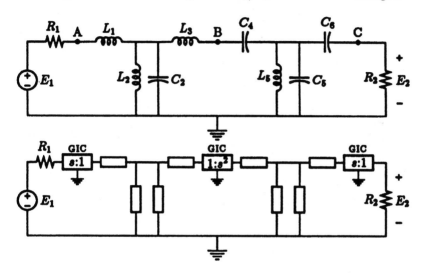

12.16 The following circuit realizes a normalized fifth-order Chebyshev lowpass filter with $\alpha_p = 0.5$ dB. Use the FDNR method to eliminate the inductors. All capacitors in the new circuit are to be 1 farad each.

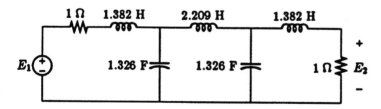

12.17 The following circuit is a third-order Chebyshev lowpass filter with $\alpha_p = 1$ dB. Simulate it using the leap-frog method.

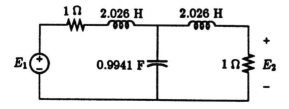

12.18 Use the leap-frog method to simulate the following LC ladder. Limit the number of op amps to four. (Hint: Apply Norton's theorem to E_1 and R_1.)

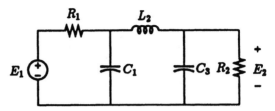

12.19 Simulate the bandpass filter circuit of Problem 12.12 by using the leap-frog method using three biquads. It is sufficient to specify the transfer function of each biquad.

12.20 Repeat the previous problem using only integrators.

Chapter 13

Switched-capacitor filters

The filters we have studied in the two previous chapters are primarily based on the assumption that they are to be implemented in discrete or hybrid forms - thick-film or thin-film. Although the op amp may be fabricated in integrated-circuit (IC) form (sometimes in groups), it is actually treated as a separate entity for our purposes. Although some of the circuits we have studied can be fabricated in IC form, this is not a routine matter. There are many attendant problems and special details that need to be addressed and overcome.

There are several alternatives in performing the filtering function using IC technology. One of them is the use of charge transfer devices. Another is surface-acoustic-wave filters. Still another approach is to use operational transconductance amplifiers. Then there is the large area of digital filters. Most of these filters require very different background knowledge than what we have at our disposal at this point. Each of these alternatives is a specialty in itself and is obviously beyond the scope of this text.

There is one type of IC filters that bears a close resemblance to filters that we have studied. They are the switched-capacitor filters. Although to study this type of filters thoroughly and in detail will require quite a different mathematical approach, the way some of these filters work can be understood based on what we have studied. We will limit our study to adapting some of the filters described earlier in this volume that are suitable for switched-capacitor circuit implementation.

13.1   An introduction

In switched-capacitor filters, a number of capacitors are switched back and forth periodically among a number of terminals. They are most suitable for implementation in metal-oxide semiconductor (MOS) form. As a consequence, switched-capacitor filters have several distinct features [BGH, HBG].

1. In MOS technology, resistors typically take up too much of the chip area. It has been found that resistors can be replaced by periodically switched capacitors. The switches are MOS transistors controlled by their gate voltages.

2. In many situations, the performance of switched-capacitor filters depends on the ratios of capacitances, not on their absolute values. Although the absolute values of the capacitance are not easily controlled in the fabrication process, their ratios can be controlled quite closely, typically to within 0.1%.

3. Capacitances typically vary in the range of 0.1 pF to 100 pF.

4. MOS switches are controlled by a clock whose frequency is typically in the range of 100 kHz to 2 MHz. Since the signal frequencies should be considerably less than the clock frequency, switched-capacitor filters are most suitable for voice frequency applications.

5. Switched-capacitor filter systems are sampled-data systems. Therefore, during each switching half-cycle, the signals are assumed to be constant. If the clock frequency, f_c, and the highest signal frequency, f_0, are too close together, aliasing may occur (if $f_c < 2f_0$) and spurious high-frequency signals will be generated. The higher the f_c/f_0 ratio is, the more likely the filter will perform as predicted. Typical f_c/f_0 ratios are in the range of 20 to 200.

6. Switched capacitors are typically realized on the same chip with other devices such as fixed capacitors, op amps, digital circuits, oscillators, as well as the remainder of the entire system unrelated to the filter. They are typically part of a large-scale integration (LSI) or very large-scale integration (VLSI) system.

13.2   Simulation of resistors by switched capacitors

We shall now look at a simple arrangement to show how a switched capacitor can simulate a resistor. Shown in Fig. 13.1(a) is a capacitor with capacitance C_R that is switched periodically between terminals 1

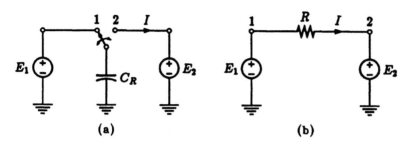

(a)   (b)

Figure 13.1: Simulation of a resistance by a switched capacitor.

and 2. When C_R is connected to 1, it acquires a charge of $C_R E_1$. When it is connected to 2, the amount of charge changes to $C_R E_2$. Hence there is a charge of $C_R(E_1 - E_2)$ delivered to E_2. If the switching occurs f_c times per second, the total charge delivered to E_2 is $f_c C_R(E_1 - E_2)$ per second, which is the value of the average current I. Hence the arrangement of Fig. 13.1(a) is equivalent to (approximately) the situation of Fig. 13.1(b) in which

$$R = \frac{E_1 - E_2}{I} = \frac{1}{f_c C_R} \tag{13.1}$$

In practice, the switching arrangement in Fig. 13.1(a) is implemented by using two MOS transistor switches as shown in Fig. 13.2. The gates of these two switches are driven by two nonoverlapping two-phase clock pulses, in which $f_c = 1/T$. When the gate voltage of a transistor is high (typically a few volts), the switch is *on*. In this state, the channel resistance is of the order of 10 kΩ. When the gate voltage is low (typically a small fraction of a volt), the switch is *off*. The channel resistance is of the order of 100 MΩ. Using the typical resistance value when the transistor is on and a capacitance of 1 pF, the time constant for charging is $\tau = 10 \times 10^3 \times 10^{-12} = 10$ nanoseconds. We must assume that this typical time constant is negligibly small compared with the signal variation. This would be assured if the highest signal frequency were much lower than $1/\tau$.

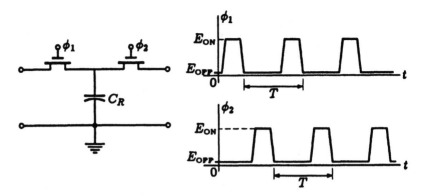

Figure 13.2: MOS implementation of the arrangement of Fig. 13.1(a).

Parasitics-insensitive circuits

In MOS technology, the capacitor C_R is always accompanied by several parasitic capacitances due to the close proximity of connecting terminals, and the small dimensions of the device layers [Ma]. Figure 13.3 shows conceptually the structure of the cross section of an MOS capacitor. As a first-order approximation, the parasitics associated with an MOS capacitor can be represented by the capacitive π network shown. The capacitance C_p is made up chiefly of the capacitances between the

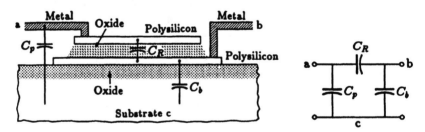

Figure 13.3: Effective parasitic capacitances associated with C_R.

top plate, the terminal connecting metal, the routing metal, and the substrate. This capacitance is typically 1% of C_R. The capacitance C_b is primarily the capacitance between the bottom plate and the substrate. It is typically 10% of C_R.

The switching arrangement of Fig. 13.4(a) renders an equivalent resistance to the arrangement in Fig. 13.1(a). Fig. 13.4(b) is the MOS implementation of the circuit in Fig. 13.4(a). When the switches are in the top position (ϕ_1 high), C_R is charged to $C_R E_R = C_R(E_1 - E_2)$. At the same time C_p is charged to $C_p E_1$ and C_b to $C_b E_2$ respectively.

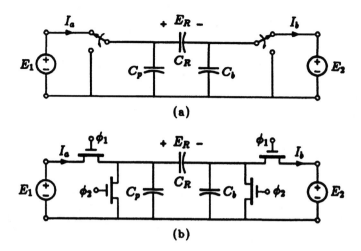

Figure 13.4: Parasitics-insensitive arrangement of a switched capacitor.

When the switches are toggled to the lower position (ϕ_2 high), all three capacitors are discharged. The periodic charging and discharging of C_R results in the average value for $I_a = I_b$ being $f_c C_R(E_1 - E_2)$. (Although I_a and I_b also contain component currents equal to $f_c C_p E_1$ and $f_c C_b E_2$ respectively, these currents are entirely local to the two voltage sources E_1 and E_2. These local currents do not contribute to the charge transfer between E_1 and E_2.) The same resistance indicated in Fig. 13.1(b) and (13.1) is simulated. Hence the net effects of the circuits in Fig. 13.1(a) and Fig. 13.4(a) are exactly the same. In the remainder of this chapter, the arrangement of Fig. 13.1(a) will frequently be used in place of that of Fig. 13.4(a) to avoid cluttering up the circuit diagram.

A slight modification of the circuit of Fig. 13.4(a) leads to a very different result. We simply reverse the phases of the two switches of Fig. 13.4(a) and obtain the circuit shown in Fig. 13.5(a). This is accomplished simply by interchanging the clocking pulses of one set of MOS switches shown in Fig. 13.5(b). Now when ϕ_1 is high, $E_R = E_1$ and C_R is charged to $C_R E_1$. Hence the average value of I_a is $f_c C_R E_1$. When ϕ_2 is high, C_R is charged to $C_R E_2$. Because of the polarity of the charge in C_R, the average value of I_b is now $-C_R f_c(E_1 + E_2)$. (Similar to the circuits in Fig. 13.4, the charging and discharging of C_p and C_b still take place. Both I_a and I_b also contain component currents caused by the charging and discharging of these capacitors. However, these charges are local to E_1 and E_2 and have no effect on the charge transfer from E_1 to E_2.) Ignoring the current contributions of C_p and C_b, the equivalent circuit of the arrangements of Fig. 13.5 is shown in Fig. 13.6.

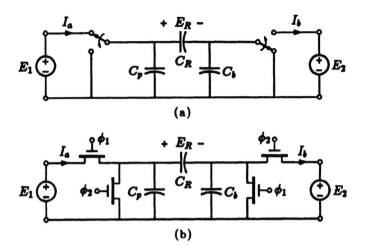

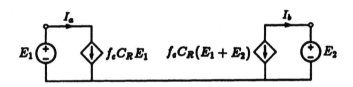

Figure 13.5: A circuit that simulates a negative resistance.

Figure 13.6: Equivalent circuit of the arrangements of Fig. 13.5.

If we consider the voltage source E_1 to be ideal, then the equivalent circuit of Fig. 13.6, in which $R = 1/(f_cC_R)$, can be simplified to that shown in Fig. 13.7(a). Although this equivalent circuit gives a wrong

current for I_a, the average value of I_b is preserved. Hence, this equivalent circuit correctly represents the charge transfer between E_1 and E_2.

If $E_2 \equiv 0$, as will be the case when the output is either grounded or connected to a virtual ground, the circuit in Fig. 13.7(a) can be represented by a negative resistance as shown in Fig. 13.7(b). Again, this equivalent circuit is only correct in so far as the average of current I_b is concerned.

13.3   Simple basic circuits

We shall now describe several basic circuits using capacitors, switches, and op amps. These circuits are the basic building blocks of switched-capacitor filters. Some of them will be used later in this chapter in the formation of switched-capacitor filters.

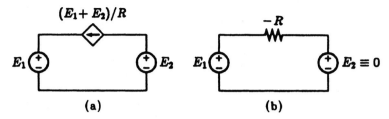

Figure 13.7: Simplified equivalent circuits of the arrangements of Fig. 13.5

13.3.1 All-capacitor op amp circuits

In Section 9.3, we compiled several basic circuits that are used in continuous-time active filters. All circuits in that section that use only resistors can be directly applied to MOS circuits by replacing every resistor with a capacitor. The input-output relationships in those circuits are directly applicable if we replace every resistance of R_i ohms with a capacitance equal to $1/R_i$ farads. For example, if we apply this idea to the inverting voltage amplifier of Fig. 9.5, we obtain the circuit of Fig. 13.8, in which

$$sC_1 E_1 + sC_2 E_2 = 0$$

Hence

$$\frac{E_2}{E_1} = -\frac{C_1}{C_2} \tag{13.2}$$

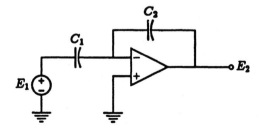

Figure 13.8: An inverting amplifier using capacitors.

A comparison of this relationship with (9.5) shows how the R's are replaced with C's.

If we apply the same idea to the inverting weighted summer of Fig. 9.8, we obtain its capacitive counterpart as shown in Fig. 13.9, in which

$$E_o = -\frac{C_1}{C_0}E_1 - \frac{C_2}{C_0}E_2 - \frac{C_3}{C_0}E_3 \tag{13.3}$$

Equation (13.3) is comparable to (9.8).

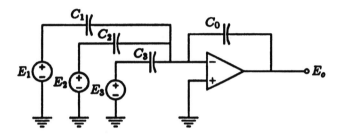

Figure 13.9: An inverting weighted summer using capacitors.

13.3.2 The inverting integrator

We replace the input resistor of Fig.9.9 with a switched capacitor as represented by the circuit of Fig. 13.1(a). We have the switched-capacitor version of that circuit and it is shown in Fig. 13.10. From (9.9) and (13.1), we have

$$\frac{E_2}{E_1} = -\frac{1}{\dfrac{1}{f_cC_1} \times C_2s} = -f_c\frac{C_1}{C_2}\cdot\frac{1}{s} \tag{13.4}$$

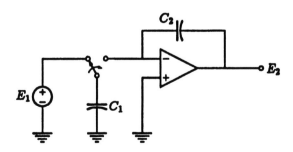

Figure 13.10: The switched-capacitor inverting integrator.

13.3.3 The inverting lossy integrator

Similarly, we replace the two resistors in the circuit of Fig.9.10 with two switched capacitors. We obtain the switched-capacitor version of the circuit and it is shown in Fig. 13.11, for which

$$\frac{E_2}{E_1} = -\frac{f_c\frac{C_1}{C_2}}{s + f_c\frac{C_3}{C_2}} = -\frac{f_cC_1}{sC_2 + f_cC_3} \tag{13.5}$$

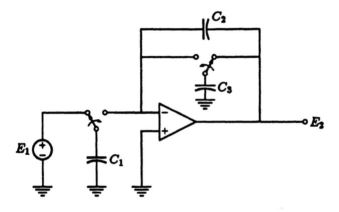

Figure 13.11: The inverting lossy integrator.

Attention should be called to the phasing of the two switches. If the phase of one of them is reversed, the circuit will perform slightly differently. To analyze circuits similar to this one will require that we study how charges are redistributed during each half-cycle. However, if $f_c \gg f_0$, then the difference of the two phasings become less noticeable. This remark will apply to many circuits in the remainder of this chapter. We will not mention this repeatedly, unless it's particularly appropriate with certain specific situations.

13.3.4 The inverting weighted summing integrator

We take the circuit of Fig. 9.11 and make similar replacements. We thus obtain the switched-capacitor version of that circuit. The switched-capacitor inverting weighted summing integrator is shown in Fig. 13.12, for which

$$E_o = -f_c\frac{C_1}{C_0}\cdot\frac{1}{s}E_1 - f_c\frac{C_2}{C_0}\cdot\frac{1}{s}E_2 - f_c\frac{C_3}{C_0}\cdot\frac{1}{s}E_3 \tag{13.6}$$

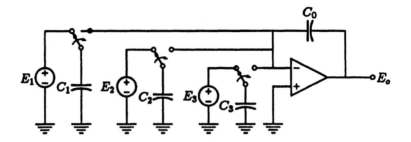

Figure 13.12: The inverting weighted summing integrator.

13.3.5 A noninverting integrator

In continuous-time operations, it is comparatively difficult to realize a noninverting integrator. We frequently cascade an inverter and a inverting integrator to obtain a noninverting integrator. An alternative to that arrangement is the circuit of Problem 9.3. In switched-capacitor circuits, negative-valued resistors are available (Fig. 13.7). Therefore, the noninverting integrator is relatively easy to realize. We simply replace the input resistor of the circuit of Fig. 9.5 with the negative resistor of Fig. 13.5(a). The result is the circuit of Fig. 13.13, for which

$$\frac{E_2}{E_1} = f_c \frac{C_1}{C_2} \cdot \frac{1}{s} \tag{13.7}$$

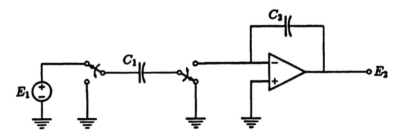

Figure 13.13: A noninverting integrator.

13.3.6 Inverting summer-integrator combination

Using the superposition principle, we could combine the circuits of Figs 13.7 and 13.9 to obtain a circuit that will perform the summing and integrating of several signals. A circuit that sums two signals and integrates a third one is shown in Fig. 13.14, in which

$$E_o = -\frac{C_1}{C_0}E_1 - \frac{C_2}{C_0}E_2 - f_c\frac{C_3}{C_0} \cdot \frac{1}{s} \cdot E_3 \qquad (13.8)$$

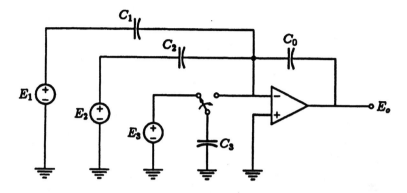

Figure 13.14: An inverting summing and integrating circuit.

13.3.7 The differential integrator

To obtain the integral of the difference of two signals in continuous-time mode is quite cumbersome. We either use an inverter to invert one signal and add the inverter output to the other, or use a differential-input op amp. The differential input to an op amp is particularly simple to obtain using the switched-capacitor arrangement. We sample the difference of two voltages and then deliver the charge to the feedback capacitor as shown in Fig. 13.15, in which

$$E_o = -f_c\frac{C_1}{C_2} \cdot \frac{1}{s}(E_1 - E_2) \qquad (13.9)$$

13.3.8 The differential lossy integrator

By adding the switched-capacitor equivalent of a resistor in parallel with the feedback capacitor C_2 of the differential integrator of Fig. 13.15, we obtain a differential-input lossy integrator as shown in Fig. 13.16. It is easy to show that

$$E_o = -\frac{f_c\dfrac{C_1}{C_2}}{s + f_c\dfrac{C_3}{C_2}}(E_1 - E_2) \qquad (13.10)$$

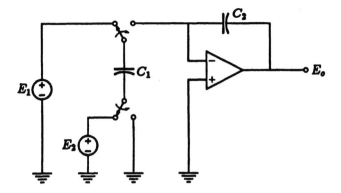

Figure 13.15: The differential integrator.

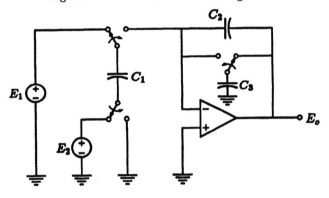

Figure 13.16: The differential lossy integrator.

13.3.9 The differential amplifier

An additional switch connected in parallel with the feedback capacitor C_2 of the differential integrator circuit of Fig. 13.15 will remove the integrating effect of C_2 and result in a differential amplifier circuit as shown in Fig. 13.17. When the switches are in the positions shown, C_1 is charged to $C_1(E_2 - E_1)$ and C_2 is empty. After the switches are toggled, C_2 is charged to $C_1(E_2 - E_1)$. Hence

$$E_o = \frac{C_1}{C_2}(E_1 - E_2) \tag{13.11}$$

This circuit is another example of the importance of the phasing of the switches. Obviously, if the phasing of S_3 is reversed from what is shown in Fig. 13.17, C_2 will be short-circuited after S_1 and S_2 have been switched away from E_1 and E_2. Then C_2 will remain uncharged and E_o

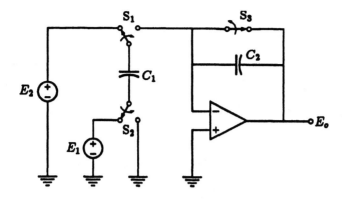

Figure 13.17: The differential amplifier.

will always be zero.

In addition, in the arrangement of Fig. 13.17, it is clear that E_o is zero when S_1 and S_2 are connected to E_1 and E_2. In asserting (13.11), it is implicit that the output voltage E_o is sampled only when S_3 is open.

If $E_2 \equiv 0$, then the circuit reduces to a noninverting amplifier. This special case of the differential amplifier is redrawn and shown in Fig. 13.18.

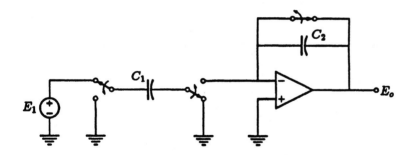

Figure 13.18: A noninverting amplifier.

13.3.10 The first-order section

Based on the first-order section of Fig. 9.13, we obtain the switched-capacitor version of the inverting first-order section by replacing each resistor with a switched capacitor. The circuit is shown in Fig. 13.19. Using the equivalent resistances of the switched capacitors C_3 and C_4, we obtain the transfer function of the first-order section

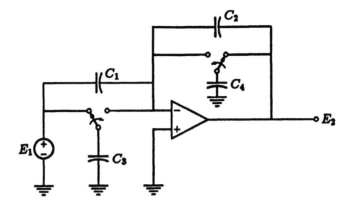

Figure 13.19: An inverting first-order section.

$$\frac{E_2}{E_1} = -\frac{C_1 s + f_c C_3}{C_2 s + f_c C_4} \tag{13.12}$$

Note the phases of the two switches connected to C_3 and C_4. They appear to be in opposite phase. Actually this arrangement causes the charging of C_3 by E_1 and the discharge of C_4 to E_2 to occur at the same time.

EXAMPLE 13.1. Design a first-order switched-capacitor filter section to realize

$$\frac{E_2}{E_1} = -\frac{s + 10^4}{s + 10^3}$$

The clock frequency is 10 kHz.

SOLUTION   From (13.12), we need $C_1 = C_2$. Choose $C_1 = C_2 = 10$ pF for the circuit of Fig. 13.19. From (13.12), we need

$$f_c \frac{C_3}{C_1} = 10^4 \qquad \Longrightarrow \qquad C_3 = 10 \text{ pF}$$

$$f_c \frac{C_4}{C_1} = 10^3 \qquad \Longrightarrow \qquad C_4 = 1 \text{ pF}$$

13.4   Switched-capacitor biquads

At this point, the reader may get the impression that any filter circuit developed on a continuous-time basis can be implemented in switched-capacitor form simply by replacing each resistor with a switched capacitor. This is not the case. Many biquad circuits adapted for switched-capacitor implementation by replacing each resistor with a switched capacitor will not perform properly. The phasing of various switches can have serious consequences when the capacitors begin to interact with one another during each half of the switching cycle.

On the other hand, certain continuous-time biquads lend themselves very conveniently to the switched-capacitor adaptation. Since we are approaching this subject by replacing continuous-time circuit components with their switched-capacitor counterparts, we shall limit our scope to the block-by-block replacement approach.

The circuit of the KHN biquad of Fig. 10.19 is reproduced in Fig. 13.20. As is explained in Section 10.3.1, this circuit is made up of three building blocks. Block N_1 is a weighted differential summer for which

$$E_4 = \frac{R_4(R_5 + R_6)}{R_5(R_3 + R_4)}E_1 - \frac{R_6}{R_5}E_2 + \frac{R_3(R_5 + R_6)}{R_5(R_3 + R_4)}E_3$$

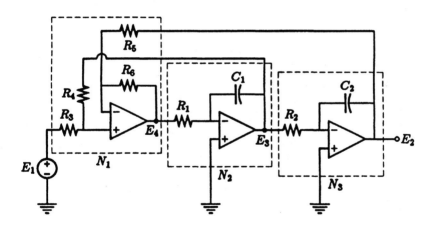

Figure 13.20: The continuous-time KHN biquad.

Since this is an all-resistor op amp circuit, we can simply replace each resistor with a capacitor in such a way that $C_i \propto 1/R_i$, $i = 3, 4, 5, 6$, for the MOS application. Blocks N_2 and N_3 are each an inverting integrator, for which

$$\frac{E_3}{E_4} = -\frac{1}{R_1 C_1 s}$$

$$\frac{E_2}{E_3} = -\frac{1}{R_2 C_2 s}$$

Hence, we simply replace R_1 and R_2 with their switched-capacitor substitute by making $C_{R_1} = 1/f_c R_1$ and $C_{R_2} = 1/f_c R_2$. The result is the circuit of Fig. 13.21.

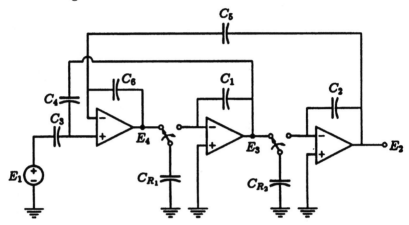

Figure 13.21: The switched-capacitor KHN biquad.

Another circuit that is amenable to block-by-block replacement is the Tow-Thomas biquad of Fig. 10.21 which is reproduced in Fig. 13.22. This circuit can be viewed as being made up of only two building blocks. Block N_1 is an inverting weighted summing integrator and

$$E_3 = -\frac{1}{R_1 C_1 s} E_3 - \frac{1}{R_4 C_1 s} E_1 - \frac{1}{R_3 C_1 s} E_2 \qquad (13.13)$$

The switched-capacitor circuit with this relationship is described in Section 13.3.4 (Fig. 13.12).

Block N_2 is a noninverting integrator and

$$E_2 = \frac{1}{R_2 C_2 s} E_3 \qquad (13.14)$$

The switched-capacitor circuit that realizes such a relationship is given in Section 13.3.5 (Fig. 13.13).

Replacing N_1 and N_2 with their switched-capacitor counterparts results in the switched-capacitor Two-Thomas biquad of Fig. 13.23. Hence, all we have to do is to take a Tow-Thomas biquad that has already been designed in continuous-time mode and replace it with the circuit of Fig. 13.23 and make $C_{R_i} = 1/f_c R_i$, $i = 1, 2, 3, 4$.

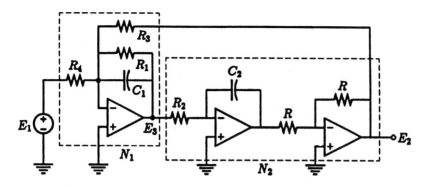

Figure 13.22: The continuous-time Tow-Thomas biquad.

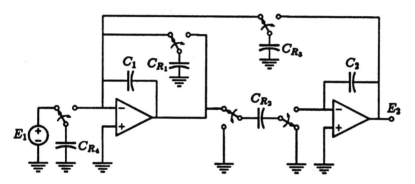

Figure 13.23: Switched-capacitor Tow-Thomas biquad.

13.5 Functional simulation of LC ladders

Another group of active filters that are easily adaptable to switched-capacitor implementation is the RC-op amp functional simulation of LC ladders. Suppose we want to simulate the doubly-terminated LC ladder of Fig. 12.18 which is reproduced in Fig. 13.24.

The input-output relationship of this ladder is simulated by the block diagram of Fig. 12.22(a) which is reproduced in Fig. 12.25. The transfer functions of the blocks are given in (12.26) which are reproduced as (13.15).

$$f_1(s) = \frac{1}{R_1 + sL_2} = \frac{1/L_2}{s + R_1/L_2} \qquad f_2(s) = \frac{1}{sC_3}$$

$$f_3(s) = \frac{1}{sL_4} \qquad f_4(s) = \frac{1}{sC_5} \tag{13.15}$$

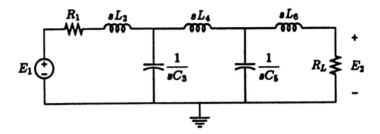

Figure 13.24: A doubly-terminated LC ladder.

$$f_5(s) = \frac{1}{sL_6} \qquad f_6(s) = R_L$$

In Section 12.5, the block diagram of Fig. 12.22(a) is modified to reduce the number of inverters required because it is not convenient to perform substraction in continuous-time circuits. In the switched-capacitor implementation, subtraction is just as easy to implement as addition. In fact, in this situation, it is preferable to implement the block diagram of Fig. 12.22(a) rather than the modified one in Fig. 12.24.

The block diagram of Fig. 12.25 is redrawn with the blocks in vertical positions in Fig. 12.26. The transfer function of each block is indicated on the diagram.

The first block in conjunction with the subtractor is a differential lossy integrator. The circuit for this device is given in Fig. 13.16. The next four blocks together with their respective associated subtractors are each a differential integrator. The circuit for such a device is given in Fig. 13.15. The last block is a noninverting amplifier. The circuit of this device is given in Fig. 13.18. Thus the circuit of Fig. 13.6 is the switched-capacitor implementation of the block diagram of Fig. 12.26.

13.6 Summary

In this chapter, we have introduced the reader to a special class of active filters that uses only switched capacitors, fixed capacitors, and op amps. These filters are suitable for MOS implementation and are used primarily in the voice-frequency range. The approach used is first to present several switched-capacitor building blocks whose operations are reasonably simple and can be followed or understood without sophisticated new analysis techniques. Then we take a few example filter circuits in the continuous-time domain, identity certain blocks and replace them with their switched-capacitor counterparts.

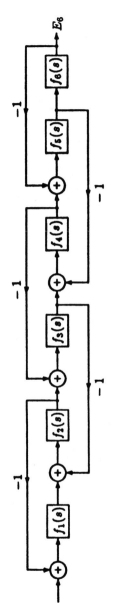

Figure 13.25: The block diagram of the system that simulates the LC ladder filter of Fig. 13.24.

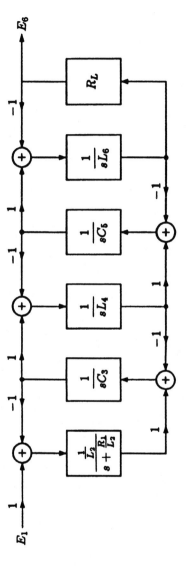

Figure 13.26: Block diagram of Fig. 13.24 redrawn.

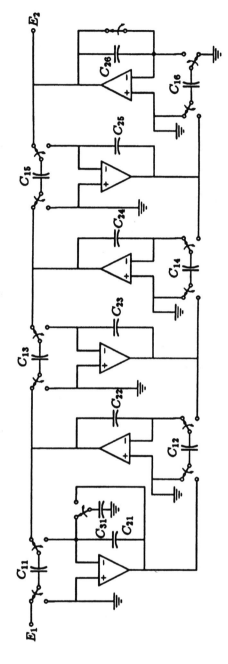

Figure 13.27: Switched-capacitor circuit implementation of the block diagram of Fig. 13.26.

This approach is, of course, very primitive and limited. But for the objective of this text, it is expedient and adequate. It is a qualitative approach. Several important assumptions are implicit in this chapter. We assume that the clock frequency is much higher than the highest signal frequency. We also tacitly assume that the op amps are ideal as we have since Chapter 9. One implication of this assumption in this application is that the slew rates of the op amps are sufficiently high that each switching can be considered to have been completed before the following sampling starts.

Also, the reader should not get the impression that every switched-capacitor filter has a continuous-time counterpart. Some switched-capacitor filters are derived directly using other suitable methods or by emulating other types of filters. They may or may not have any continuous-time counterparts.

To deal with this class of filters precisely we need to take into account the effect of switching - mainly what is happening during each half of a switching cycle. Since filters of this type are really sampled-data systems, the z-transform is the primary tool to use to analyze these filters quantitatively. We will not attempt to do this here.

Problems

13.1 This problem illustrates the importance of the phasing of the switches in a circuit. Both circuits appear to be equivalent to two resistors, $1/f_cC_1$ and $1/f_cC_2$, connected in series. Carefully examine the equivalent I in each circuit and determine the correct equivalent resistance between E_1 and E_2.

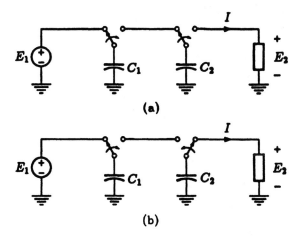

13.2 Determine the equivalent resistance between E_1 and E_2 in the following circuit.

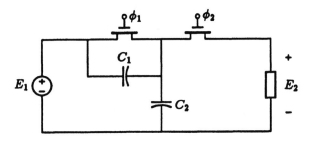

13.3 Obtain the voltage transfer function, E_2/E_1, of the following circuit.

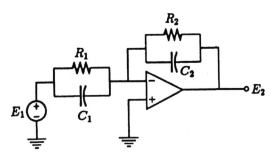

13.4 Based on the following circuit, obtain a switched-capacitor circuit to realize

$$\frac{E_2}{E_1} = -\frac{3s + 10^3}{s + 500}$$

Assume $f_c = 100$ kHz. The largest capacitance is to be 20 pF.

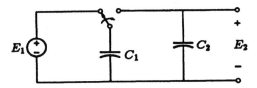

13.5 First obtain the voltage transfer function, E_2/E_1, of the following circuit. Then implement it using only capacitors, op amps, and switches. The clock frequency is 250 kHz.

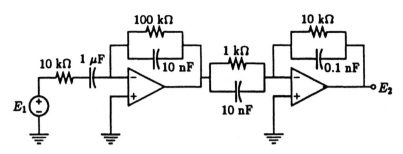

13.6 In the following circuit, $C_1 = C_2 = 0.5$ nF, $C_3 = C_4 = 0.25$ nF, $C_5 = 5$ pF, $C_6 = 2$ nF, and $f_c = 200$ kHz. Obtain the voltage transfer function, E_2/E_1.

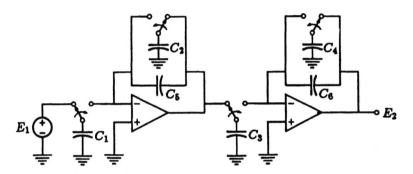

13.7 By summing the average currents at the node connected to the inverting terminal of the op amp, derive the voltage transfer function (13.10) of the differential lossy integrator of Fig. 13.16.

13.8 Design a switched-capacitor KHN biquad to realize

$$\frac{E_2}{E_1} = \frac{10^8}{s^2 + 10^4 s + 10^8}$$

Let $f_c = 100$ kHz and let all feedback capacitors be 10 pF.

13.9 Repeat the previous problem using a Tow-Thomas biquad.

13.10 The circuit below is a two-integrator biquad suitable for switched-capacitor implementation. (a) Obtain the voltage transfer function, E_2/E_1. (b) Implement this circuit using only capacitors, switches, and op amps. Use as few circuit elements as possible.

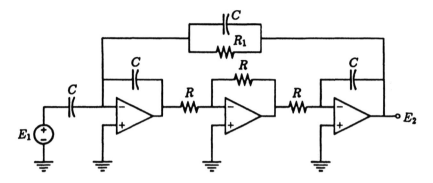

13.11 Use the leap-frog method to form a switched-capacitor circuit to simulate the following filter, which is a normalized lowpass Chebyshev filter with $\alpha_p = 0.1$ dB.

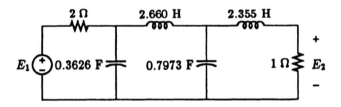

13.12 In the circuit of Fig. 13.6, assume each fixed op amp feedback capacitance to be 1 F. Obtain the expressions for the values of all capacitances in terms of f_c and the element values in the LC ladder.

13.13 Design a fourth-order switched-capacitor lowpass filter with a Chebyshev characteristic and $\alpha_p = 0.1$ dB, $\omega_p = 5,000$ rad/sec, and $f_c = 100$ kHz.

(a) Use a cascade of two KHN biquads. Make the dc gain of each biquad unity. Let all unswitched capacitances be 1 pF.

(b) Use a cascade of two Tow-Thomas biquads.

Appendix A
Tables of filter functions

Table A.1. Butterworth polynomials in factored form

n	Polynomials
1	$s + 1$
2	$s^2 + 1.414214s + 1$
3	$(s^2 + s + 1)(s + 1)$
4	$(s^2 + 0.765367s + 1)(s^2 + 1.847759s + 1)$
5	$(s^2 + 0.618034s + 1)(s^2 + 1.618034s + 1)(s + 1)$
6	$(s^2 + 0.517638s + 1)(s^2 + 1.414214s + 1)(s^2 + 1.931852s + 1)$
7	$(s^2 + 0.445042s + 1)(s^2 + 1.246980s + 1)(s^2 + 1.801938s + 1)(s + 1)$
8	$(s^2 + 0.390181s + 1)(s^2 + 1.111140s + 1)(s^2 + 1.662939s + 1)(s^2 + 1.961571s + 1)$
9	$(s^2 + 0.347296s + 1)(s^2 + s + 1)(s^2 + 1.532089s + 1)(s^2 + 1.879385s + 1)(s + 1)$
10	$(s^2 + 0.312869s + 1)(s^2 + 0.907981s + 1)(s^2 + 1.414214s + 1)(s^2 + 1.782013s + 1)$ $\times (s^2 + 1.975377s + 1)$

Table A.2. Butterworth polynomials in expanded form

n	Polynomials
1	$s + 1$
2	$s^2 + 1.414214s + 1$
3	$s^3 + 2s^2 + 2s + 1$
4	$s^4 + 2.613126s^3 + 3.414214s^2 + 2.613126s + 1$
5	$s^5 + 3.236068s^4 + 5.236068s^3 + 5.236068s^2 + 3.236068s + 1$
6	$s^6 + 3.863703s^5 + 7.464102s^4 + 9.141620s^3 + 7.464102s^2 + 3.863703s + 1$
7	$s^7 + 4.493959s^6 + 10.097835s^5 + 14.591794s^4 + 14.591794s^3 + 10.097835s^2 + 4.493959s + 1$
8	$s^8 + 5.125831s^7 + 13.137071s^6 + 21.846151s^5 + 25.688356s^4 + 21.846151s^3 + 13.137071s^2 + 5.125831s + 1$
9	$s^9 + 5.758770s^8 + 16.581719s^7 + 31.163437s^6 + 41.986386s^5 + 41.986386s^4 + 31.163437s^3 + 16.581719s^2$ $+ 5.758770s + 1$
10	$s^{10} + 6.392453s^9 + 20.431729s^8 + 42.802061s^7 + 64.882396s^6 + 74.233429s^5 + 64.882396s^4 + 42.802061s^3$ $+ 20.431729s^2 + 6.392453s + 1$

Table A.3. Denominator polynomials in factored form for Chebyshev filters

$$\alpha_p = 0.1 \text{ dB} \quad (\epsilon^2 = 0.0232930 \quad \epsilon = 0.152620)$$

n	Polynomials
1	$s + 6.552203$
2	$s^2 + 2.372356s + 3.314037$
3	$(s^2 + 0.969406s + 1.689747)(s + 0.969406)$
4	$(s^2 + 0.528313s + 1.330031)(s^2 + 1.275460s + 0.622925)$
5	$(s^2 + 0.333067s + 1.194937)(s^2 + 0.871982s + 0.635920)(s + 0.538914)$
6	$(s^2 + 0.229387s + 1.129387)(s^2 + 0.626696s + 0.696374)(s^2 + 0.856083s + 0.263361)$
7	$(s^2 + 0.167682s + 1.092446)(s^2 + 0.469834s + 0.753222)(s^2 + 0.678930s + 0.330217)(s + 0.376778)$
8	$(s^2 + 0.127960s + 1.069492)(s^2 + 0.364400s + 0.798894)(s^2 + 0.545363s + 0.416210)$ $(s^2 + 0.643300s + 0.145612)$
9	$(s^2 + 0.100876s + 1.054214)(s^2 + 0.290461s + 0.834368)(s^2 + 0.445012s + 0.497544)$ $(s^2 + 0.545888s + 0.201345)(s + 0.290461)$
10	$(s^2 + 0.0815773s + 1.043513)(s^2 + 0.236747 + 0.861878)(s^2 + 0.368742s + 0.567985)$ $(s^2 + 0.464642s + 0.274093)(s^2 + 0.515059s + 0.0924569)$

Table A.4. Denominator polynomials in expanded form for Chebyshev filters

$$\alpha_p = 0.1 \text{ dB} \quad (\epsilon^2 = 0.0232930 \quad \epsilon = 0.152620)$$

n	Polynomials
1	$s + 6.552203$
2	$s^2 + 2.372356s + 3.314037$
3	$s^3 + 1.938811s^2 + 2.629495s + 1.638051$
4	$s^4 + 1.803773s^3 + 2.626798s^2 + 2.025501s + 0.828509$
5	$s^5 + 1.743963s^4 + 2.770704s^3 + 2.396959s^2 + 1.435558s + 0.409513$
6	$s^6 + 1.712166s^5 + 2.965756s^4 + 2.779050s^3 + 2.047841s^2 + 0.901760s + 0.207127$
7	$s^7 + 1.693224s^6 + 3.183504s^5 + 3.169246s^4 + 2.705144s^3 + 1.482934s^2 + 0.561786s + 0.102378$
8	$s^8 + 1.681023s^7 + 3.412919s^6 + 3.564770s^5 + 3.418452s^4 + 2.159241s^3 + 1.066626s^2 + 0.326431s + 0.0517818$
9	$s^9 + 1.672699s^8 + 3.648961s^7 + 3.963845s^6 + 4.191611s^5 + 2.933873s^4 + 1.734120s^3 + 0.694211s^2 + 0.191760s + 0.0255945$
10	$s^{10} + 1.666766s^9 + 3.889055s^8 + 4.365370s^7 + 5.026177s^6 + 3.808504s^5 + 2.579035s^4 + 1.229664s^3 + 0.457216s^2 + 0.107034s + 0.0129455$

Table A.5. Denominator polynomials in factored form for Chebyshev filters

$$\alpha_p = 0.5 \text{ dB} \quad (\epsilon^2 = 0.122018 \quad \epsilon = 0.349311)$$

n	Polynomials
1	$s + 2.862775$
2	$s^2 + 1.425625s + 1.516203$
3	$(s^2 + 0.626456s + 1.142448)(s + 0.626456)$
4	$(s^2 + 0.350706s + 1.063519)(s^2 + 0.846680s + 0.356412)$
5	$(s^2 + 0.223926s + 1.035784)(s^2 + 0.586245s + 0.476767)(s + 0.362320)$
6	$(s^2 + 0.155300s + 1.023023)(s^2 + 0.424288s + 0.590010)(s^2 + 0.579588s + 0.156997)$
7	$(s^2 + 0.114006s + 1.016108)(s^2 + 0.319439s + 0.676884)(s^2 + 0.461602s + 0.253878)(s + 0.256170)$
8	$(s^2 + 0.0872402s + 1.011932)(s^2 + 0.248439s + 0.741334)(s^2 + 0.371815s + 0.358650)$ $(s^2 + 0.438586s + 0.0880523)$
9	$(s^2 + 0.0689054s + 1.009211)(s^2 + 0.198405s + 0.789365)(s^2 + 0.303975s + 0.452541)$ $(s^2 + 0.372880s + 0.156342)(s + 0.198405)$
10	$(s^2 + 0.0557988s + 1.007335)(s^2 + 0.161934s + 0.825700)(s^2 + 0.252219s + 0.531807)$ $(s^2 + 0.317814s + 0.237915)(s^2 + 0.352300s + 0.0562789)$

Table A.6. **Denominator polynomials in expanded form for Chebyshev filters**

$$\alpha_p = 0.5 \text{ dB} \quad (\epsilon^2 = 0.122018 \quad \epsilon = 0.349311)$$

n	Polynomials
1	$s + 2.862775$
2	$s^2 + 1.425625s + 1.516203$
3	$s^3 + 1.252913s^2 + 1.534895s + 0.715694$
4	$s^4 + 1.197386s^3 + 1.716866s^2 + 1.025455s + 0.379051$
5	$s^5 + 1.172491s^4 + 1.937375s^3 + 1.309575s^2 + 0.752518s + 0.178923$
6	$s^6 + 1.159176s^5 + 2.171845s^4 + 1.589764s^3 + 1.171861s^2 + 0.432367s + 0.0947627$
7	$s^7 + 1.151218s^6 + 2.412651s^5 + 1.869408s^4 + 1.647903s^3 + 0.755651s^2 + 0.282072s + 0.0447309$
8	$s^8 + 1.146080s^7 + 2.656750s^6 + 2.149217s^5 + 2.184015s^4 + 1.148589s^3 + 0.573560s^2 + 0.152544s + 0.0236907$
9	$s^9 + 1.142571s^8 + 2.902734s^7 + 2.429330s^6 + 2.781499s^5 + 1.611388s^4 + 0.983620s^3 + 0.340819s^2$ $+ 0.0941198s + 0.0111827$
10	$s^{10} + 1.140066s^9 + 3.149876s^8 + 2.709741s^7 + 3.440927s^6 + 2.144237s^5 + 1.527431s^4 + 0.626969s^3$ $+ 0.237269s^2 + 0.0492855s + 0.00592267$

Table A.7. Denominator polynomials in factored form for Chebyshev filters

$$\alpha_p = 1\,\text{dB} \quad (\epsilon^2 = 0.258925 \quad \epsilon = 0.508847)$$

n	Polynomials
1	$s + 1.965227$
2	$s^2 + 1.097734s + 1.102510$
3	$(s^2 + 0.494171s + 0.994205)(s + 0.494171)$
4	$(s^2 + 0.279072s + 0.986505)(s^2 + 0.673739s + 0.279398)$
5	$(s^2 + 0.178917s + 0.988315)(s^2 + 0.468410s + 0.429298)(s + 0.289493)$
6	$(s^2 + 0.124362s + 0.990732)(s^2 + 0.339763s + 0.557720)(s^2 + 0.464125s + 0.124707)$
7	$(s^2 + 0.0914180s + 0.992679)(s^2 + 0.256147s + 0.653456)(s^2 + 0.370144s + 0.230450)(s + 0.205414)$
8	$(s^2 + 0.0700165s + 0.994141)(s^2 + 0.199390s + 0.723543)(s^2 + 0.298408s + 0.340859)$ $(s^2 + 0.351997s + 0.0702612)$
9	$(s^2 + 0.0553349s + 0.995233)(s^2 + 0.159330s + 0.775386)(s^2 + 0.244108s + 0.438562)$ $(s^2 + 0.299443s + 0.142364)(s + 0.159330)$
10	$(s^2 + 0.0448289s + 0.996058)(s^2 + 0.130099s + 0.814423)(s^2 + 0.202633s + 0.520530)$ $(s^2 + 0.255333s + 0.226637)(s^2 + 0.283039s + 0.0450019)$

Table A.8. Denominator polynomials in expanded form for Chebyshev filters

$$\alpha_p = 1 \text{ dB} \quad (\epsilon^2 = 0.258925 \quad \epsilon = 0.508847)$$

n	Polynomials
1	$s + 1.965227$
2	$s^2 + 1.097734s + 1.102510$
3	$s^3 + 0.988341s^2 + 1.238409s + 0.491307$
4	$s^4 + 0.952811s^3 + 1.453925s^2 + 0.742619s + 0.275628$
5	$s^5 + 0.936820s^4 + 1.688816s^3 + 0.974396s^2 + 0.580534s + 0.122827$
6	$s^6 + 0.928251s^5 + 1.930825s^4 + 1.202140s^3 + 0.939346s^2 + 0.307081s + 0.0689069$
7	$s^7 + 0.923123s^6 + 2.176078s^5 + 1.428794s^4 + 1.357545s^3 + 0.548620s^2 + 0.213671s + 0.0307067$
8	$s^8 + 0.919811s^7 + 2.423026s^6 + 1.655156s^5 + 1.836902s^4 + 0.846824s^3 + 0.447826s^2 + 0.107345s + 0.0172267$
9	$s^9 + 0.917548s^8 + 2.670947s^7 + 1.881480s^6 + 2.378119s^5 + 1.201607s^4 + 0.786311s^3 + 0.244186s^2$ $+ 0.0706048s + 0.00767667$
10	$s^{10} + 0.915932s^9 + 2.919466s^8 + 2.107852s^7 + 2.981509s^6 + 1.612986s^5 + 1.244491s^4 + 0.455389s^3$ $+ 0.182451s^2 + 0.0344971s + 0.00430668$

Table A.9. Denominator polynomials in factored form for Chebyshev filters

$$\alpha_p = 2 \text{ dB} \quad (\epsilon^2 = 0.584893 \quad \epsilon = 0.764783)$$

n	Polynomials
1	$s + 1.307560$
2	$s^2 + 0.803816s + 0.823060$
3	$(s^2 + 0.368911s + 0.886095)(s + 0.368911)$
4	$(s^2 + 0.209775s + 0.928675)(s^2 + 0.506440s + 0.221568)$
5	$(s^2 + 0.134922s + 0.952167)(s^2 + 0.353230s + 0.393150)(s + 0.218308)$
6	$(s^2 + 0.0939464s + 0.965952)(s^2 + 0.256666s + 0.532939)(s^2 + 0.350613s + 0.0999261)$
7	$(s^2 + 0.0691327s + 0.974615)(s^2 + 0.193706s + 0.635391)(s^2 + 0.279913s + 0.212386)(s + 0.155340)$
8	$(s^2 + 0.0529848s + 0.980380)(s^2 + 0.150888s + 0.709782)(s^2 + 0.225820s + 0.327099)$ $(s^2 + 0.266372s + 0.0565006)$
9	$(s^2 + 0.0418943s + 0.984398)(s^2 + 0.120630s + 0.764552)(s^2 + 0.184816s + 0.427727)$ $(s^2 + 0.226710s + 0.131529)(s + 0.120630)$
10	$(s^2 + 0.0339516s + 0.987304)(s^2 + 0.0985315s + 0.805669)(s^2 + 0.153466s + 0.511776)$ $(s^2 + 0.193379s + 0.217883)(s^2 + 0.214363s + 0.0362477)$

Table A.10. Denominator polynomials in expanded form for Chebyshev filters

$$\alpha_p = 2 \text{ dB} \quad (\epsilon^2 = 0.584893 \quad \epsilon = 0.764783)$$

n	Polynomials
1	$s + 1.307560$
2	$s^2 + 0.803816s + 0.823060$
3	$s^3 + 0.737822s^2 + 1.022190s + 0.326890$
4	$s^4 + 0.716215s^3 + 1.256482s^2 + 0.516798s + 0.205765$
5	$s^5 + 0.706461s^4 + 1.499543s^3 + 0.693477s^2 + 0.459349s + 0.0817225$
6	$s^6 + 0.701226s^5 + 1.745859s^4 + 0.867015s^3 + 0.771462s^2 + 0.210271s + 0.0514413$
7	$s^7 + 0.698091s^6 + 1.993665s^5 + 1.039546s^4 + 1.144597s^3 + 0.382638s^2 + 0.166126s + 0.0204306$
8	$s^8 + 0.696065s^7 + 2.242253s^6 + 1.211712s^5 + 1.579581s^4 + 0.598221s^3 + 0.358704s^2 + 0.0729373s + 0.0128603$
9	$s^9 + 0.694679s^8 + 2.491290s^7 + 1.383746s^6 + 2.076748s^5 + 0.856865s^4 + 0.644468s^3 + 0.168447s^2$ $+ 0.0543756s + 0.00510766$
10	$s^{10} + 0.693690s^9 + 2.740603s^8 + 1.555742s^7 + 2.636251s^6 + 1.158529s^5 + 1.038910s^4 + 0.317756s^3$ $+ 0.144006s^2 + 0.0233347s + 0.00321508$

Table A.11. Denominator polynomials in factored form for Chebyshev filters

$$\alpha_p = 3 \text{ dB} \quad (\epsilon^2 = 0.995262 \quad \epsilon = 0.997628)$$

n	Polynomials
1	$s + 1.002377$
2	$s^2 + 0.644900s + 0.707948$
3	$(s^2 + 0.298620s + 0.839174)(s + 0.298620)$
4	$(s^2 + 0.170341s + 0.903087)(s^2 + 0.411239s + 0.195980)$
5	$(s^2 + 0.109720s + 0.936025)(s^2 + 0.287250s + 0.377009)(s + 0.177530)$
6	$(s^2 + 0.0764590s + 0.954830)(s^2 + 0.208890s + 0.521818)(s^2 + 0.285349s + 0.0888048)$
7	$(s^2 + 0.0562913s + 0.966483)(s^2 + 0.157725s + 0.627259)(s^2 + 0.227919s + 0.204254)(s + 0.126485)$
8	$(s^2 + 0.0431563s + 0.974173)(s^2 + 0.122899s + 0.703575)(s^2 + 0.183931s + 0.320892)$ $(s^2 + 0.216961s + 0.0502939)$
9	$(s^2 + 0.0341304s + 0.979504)(s^2 + 0.0982746s + 0.759658)(s^2 + 0.150565s + 0.422834)$ $(s^2 + 0.184696s + 0.126636)(s + 0.0982746)$
10	$(s^2 + 0.0276639s + 0.983346)(s^2 + 0.0802838s + 0.801711)(s^2 + 0.125045s + 0.507818)$ $(s^2 + 0.157566s + 0.213926)(s^2 + 0.174663s + 0.0322899)$

Table A.12. **Denominator polynomials in expanded form for Chebyshev filters**

$$\alpha_p = 3 \text{ dB} \quad (\epsilon^2 = 0.995262 \quad \epsilon = 0.997628)$$

n	Polynomials
1	$s + 1.002377$
2	$s^2 + 0.644900s + 0.707948$
3	$s^3 + 0.597240s^2 + 0.928348s + 0.250594$
4	$s^4 + 0.581580s^3 + 1.169118s^2 + 0.404768s + 0.176987$
5	$s^5 + 0.574500s^4 + 1.415025s^3 + 0.548937s^2 + 0.407966s + 0.0626486$
6	$s^6 + 0.570698s^5 + 1.662848s^4 + 0.690610s^3 + 0.699098s^2 + 0.163430s + 0.0442467$
7	$s^7 + 0.568420s^6 + 1.911551s^5 + 0.831441s^4 + 1.051845s^3 + 0.300017s^2 + 0.146153s + 0.0156621$
8	$s^8 + 0.566948s^7 + 2.160715s^6 + 0.971947s^5 + 1.466699s^4 + 0.471899s^3 + 0.320765s^2$ $+0.0564813s + 0.0110617$
9	$s^9 + 0.565941s^8 + 2.410144s^7 + 1.112322s^6 + 1.943860s^5 + 0.678931s^4 + 0.583506s^3$ $+0.131390s^2 + 0.0475908s + 0.00391554$
10	$s^{10} + 0.565222s^9 + 2.659738s^8 + 1.252647s^7 + 2.483421s^6 + 0.921066s^5 + 0.949921s^4$ $+0.249204s^3 + 0.127756s^2 + 0.0180313s + 0.00276542$

Table A.13. Denominator polynomials in factored form for Bessel-Thomson filters

n	$y_n(s)$
1	$s + 1$
2	$s^2 + 3s + 3$
3	$(s^2 + 3.67781s + 6.45943)(s + 2.32219)$
4	$(s^2 + 4.20758s + 11.4878)(s^2 + 5.79242s + 9.140131)$
5	$(s^2 + 4.64935s + 18.1563)(s^2 + 6.70391s + 14.2725)(s + 3.64674)$
6	$(s^2 + 5.03186s + 26.5140)(s^2 + 7.47142s + 20.8528)(s^2 + 8.49672s + 18.8011)$
7	$(s^2 + 5.37135s + 36.5968)(s^2 + 8.14028s + 28.9365)(s^2 + 9.51658s + 25.6664)(s + 4.97179)$
8	$(s^2 + 5.67797s + 48.4320)(s^2 + 8.73658s + 38.5693)(s^2 + 10.4097s + 33.9347)(s^2 + 11.1758s + 31.9772)$
9	$(s^2 + 5.95852s + 62.0414)(s^2 + 9.27688s + 49.7885)(s^2 + 11.2088s + 43.6466)(s^2 + 12.2587s + 40.5893)$ $(s + 6.29702)$
10	$(s^2 + 6.21783s + 77.4427)(s^2 + 9.77244s + 62.6256)(s^2 + 11.93506s + 54.8392)(s^2 + 13.2306s + 50.5824)$ $(s^2 + 13.8441s + 48.6675)$

Table A.14. Denominator polynomials in expanded form for Bessel-Thomson filters

n	$y_n(s)$
1	$s + 1$
2	$s^2 + 3s + 3$
3	$s^3 + 6s^2 + 15s + 15$
4	$s^4 + 10s^3 + 45s^2 + 105s + 105$
5	$s^5 + 15s^4 + 105s^3 + 420s^2 + 945s + 945$
6	$s^6 + 21s^5 + 210s^4 + 1260s^3 + 4725s^2 + 10,395s + 10,395$
7	$s^7 + 28s^6 + 378s^5 + 3150s^4 + 17,325s^3 + 62,370s^2 + 135,135s + 135,135$
8	$s^8 + 36s^7 + 630s^6 + 6930s^5 + 51,975s^4 + 270,270s^3 + 945,945s^2 + 2,027,025s + 2,027,025$
9	$s^9 + 45s^8 + 990s^7 + 13,860s^6 + 135,135s^5 + 945,945s^4 + 4,729,725s^3 + 16,216,200s^2$ $+ 34,459,425s + 34,459,425$
10	$s^{10} + 55s^9 + 1485s^8 + 25,740s^7 + 315,315s^6 + 2,837,835s^5 + 18,918,900s^4 + 91,891,800s^3$ $+ 310,134,825s^2 + 654,729,075s + 654,729,075$

Table A.15. **Coefficients of elliptic-filter functions**

$\alpha_p = 1.0$ dB $\qquad \alpha_s = 30$ dB

n	ω_s	K	a_i	b_i	ω_{si}^2
2	4.00411	3.16227766E-02	1.07759259	1.11970398	31.55754595
3	1.73251	1.48990916E-01	0.41056699	1.01620588	3.81651478
			0.55955791		
4	1.25038	3.16227766E-02	0.16585179	1.00462740	1.72085490
			0.76344886	0.43519018	7.16044227
5	1.09554	1.16339651E-01	0.06761574	1.00175647	1.24800301
			0.40183952	0.69401084	2.15745109
			0.45056343		
6	1.03799	3.16227766E-02	0.02761548	1.00070851	1.09474927
			0.18308029	0.85992383	1.36625254
			0.69986747	0.36342207	5.89295642
7	1.01536	1.11944662E-01	0.01128388	1.00028890	1.03768963
			0.07825052	0.94008233	1.13586893
			0.38893445	0.64459291	1.99333326
			0.43391247		

In Tables A.15 through A.18, it is assumed that the network functions have the form

$$H(s) = \frac{K \prod_i (s^2 + \omega_{si}^2)}{(s + a_j) \prod_i (s^2 + a_i s + b_i)}$$

where $(s + a_j)$ represents the real pole when n is odd. The pass band has been normalized to unity. In other words, $\omega_p = 1$. The coefficient K has been calibrated such that

$$|H(j\omega)|_{\max} = 1.$$

Table A.16. Coefficients of elliptic-filter functions

$\alpha_p = 1.0$ dB $\alpha_s = 60$ dB

n	ω_s	K	a_i	b_i	ω_{si}^2
3	5.02128	1.49123379E-02	0.48551654	0.99675014	33.44898430
			0.50042888		
4	2.46078	1.00000000E-03	0.25602619	0.99088666	7.00407028
			0.69206751	0.30396887	38.32680339
5	1.67161	7.50482514E-03	0.14346244	0.99356610	3.02950855
			0.46147548	0.49519149	7.04420515
			0.32551786		
6	1.34354	1.00000000E-03	0.08201430	0.99607224	1.89013542
			0.29619886	0.66715645	3.01060396
			0.54291159	0.18152493	18.89331870
7	1.18547	6.22855504E-03	0.04718632	0.99769257	1.44300836
			0.18343552	0.79203961	1.88026299
			0.40767795	0.36650115	4.75831661
			0.27765730		
8	1.10308	1.00000000E-03	0.02720164	0.99866071	1.23546746
			0.11061628	0.87422646	1.43822434
			0.28004552	0.56045674	2.40805087
			0.49515512	0.15050572	15.33957836
9	1.05822	5.87530831E-03	0.01568983	0.99922575	1.12970460
			0.06554114	0.92540541	1.23303297
			0.18022043	0.71619551	1.65393510
			0.38719585	0.32870313	4.22479678
			0.26270204		
10	1.03318	1.00000000E-03	0.00905076	0.99955292	1.07288063
			0.03841152	0.95626841	1.12840768
			0.11107511	0.82488976	1.33580569
			0.27228784	0.52627289	2.24894555
			0.47955497	0.14112107	14.34976215

Table A.17. Coefficients of elliptic-filter functions

$\alpha_p = 0.5$ dB $\qquad \omega_s/\omega_p = 1.50$

n	α_s (dB)	K	a_i	b_i	ω_{si}^2
2	8.28162	3.85406679E-01	1.03153404	1.60319288	3.92705098
3	21.92313	3.14095748E-01	0.45285639	1.14916832	2.80601409
			0.76695213		
4	36.25132	1.53969335E-02	0.25496154	1.06043768	2.53555301
			0.92000559	0.47182654	12.09930946
5	50.60705	1.91969228E-02	0.16346207	1.03189397	2.42551467
			0.57023587	0.57601197	5.43764461
			0.42597073		
6	64.96381	5.64689356E-04	0.11367204	1.01961982	2.36928876
			0.38346065	0.67467294	3.92705098
			0.65350144	0.20894915	25.82724179
7	79.32060	9.85554101E-04	0.08359620	1.01330080	2.33652227
			0.27493000	0.74776048	3.31399017
			0.48846554	0.32196599	9.53007807
			0.29811730		
8	93.67739	2.07076345E-05	0.06404804	1.00963165	2.31569730
			0.20675371	0.80046127	2.99566041
			0.37105478	0.43613983	6.02182423
			0.49962799	0.11713721	45.05997715
9	108.03418	4.64671198E-05	0.05063149	1.00731182	2.30161638
			0.16120888	0.83892087	2.80601409
			0.28892242	0.53210975	4.63633606
			0.40825939	0.20202715	15.01434140
			0.22996083		
10	122.39097	7.59366277E-07	0.04102694	1.00574990	2.29164107
			0.12928217	0.86756911	2.68264144
			0.23047361	0.60896852	3.92705098
			0.33333927	0.29829739	8.75076419
			0.40306359	0.07482336	69.79149744

Table A.18. Coefficients of elliptic-filter functions

$\alpha_p = 1$ dB $\omega_s/\omega_p = 1.50$

n	α_s (dB)	K	a_i	b_i	ω_{si}^2
2	11.19387	2.75617220E-01	0.87941826	1.21443112	3.92705098
3	25.17584	2.15619224E-01	0.37539605	1.02371395	2.80601409
			0.59101528		
4	39.51826	1.05702894E-02	0.20881889	0.99881116	2.53555301
			0.72997679	0.36428121	12.09930946
5	53.87453	1.31782287E-02	0.13308128	0.99495747	2.42551467
			0.45774932	0.51706855	5.43764461
			0.33784626		
6	68.23130	3.87645779E-04	0.09223297	0.99488729	2.36928876
			0.30896048	0.63825477	3.92705098
			0.52160761	0.16459994	25.82724179
7	82.58809	6.76559336E-04	0.06768826	0.99553059	2.33652227
			0.22175674	0.72329118	3.31399017
			0.39158024	0.29109535	9.53007807
			0.23818831		
8	96.94488	1.42152962E-05	0.05178884	0.99622376	2.31569730
			0.16679963	0.78296820	2.99566041
			0.29816955	0.41383781	6.02182423
			0.40010554	0.09300754	45.05997715
9	111.30167	3.18985672E-05	0.04090141	0.99682433	2.30161638
			0.13004272	0.82581278	2.80601409
			0.23246106	0.51547885	4.63633606
			0.32756290	0.18339923	15.01434140
			0.18427506		
10	125.65847	5.21286800E-07	0.03311992	0.99731647	2.29164107
			0.10426784	0.85738256	2.68264144
			0.18555369	0.59620675	3.92705098
			0.26780552	0.28365233	8.75076419
			0.32333253	0.05963622	69.79149744

Bibliography

[AM]   D. Åkerberg and K. Mossberg, 'A versatile active RC building block with inherent compensation for the finite bandwidth of the amplifier,' *IEEE Trans. Circuits and Systems*, vol. CAS-21, pp. 75-78, Jan. 1974.

[An]   A. Antoniou, 'Gyrators using operational amplifiers,' *Electronics Letters*, vol. 3, pp. 350-352, Aug. 1967; and 'Realization of gyrators using operational amplifiers and their use in RC-active network synthesis,' *Proc. IEE*, vol. 116, pp. 1838-1850, Nov. 1969.

[Ba]   J. R. Bainter, 'Active filter has stable notch, and response can be regulated,' *Electronics*, vol. pp. 115-117, Oct. 2, 1975.

[BGH] R. W. Brodersen, P. R. Gray, and D. A. Hodges, 'MOS switched capacitor filters,' *Proc. IEEE*, vol. 67, pp. 61-75, Jan. 1979.

[Bo]   H. W. Bode, *Network Analysis and Feedback Amplifier Design*, D. Van Nostrand Co., Inc., New York, 1945.

[Br]   L. T. Bruton, 'Frequency selectivity using positive impedance converter-type networks,' *Proc. IEEE*, vol. 56, pp. 1378-1379, Aug. 1968; and 'Network transfer function using the concept of frequency dependent negative resistance,' *IEEE Trans. Circuit Theory*, vol. CT-18, pp. 406-408, Aug. 1969.

[Bu]   S. Butterworth, 'On the theory of filter amplifiers,' *Wireless Engineering*, vol. 7, pp. 536-541, Oct. 1930.

[Ca]   W. Cauer, *Siebschaltungen*, V.D.I. Verlag G.m.b.H., Berlin, 1931.

[Da]   S. Darlington, 'Synthesis of reactance 4-pole which produce prescribed insertion loss characteristics,' *Journ. Math. Phys.*, vol. 18, pp. 257-353, Sept. 1939.

[De]   T. Delyiannis, 'High-Q factor circuit with reduced sensitivity,' *Electronics Letters*, vol. 4, p. 577, Dec. 1968.

[FHH] J. J. Friend, C. A. Harris, and D. Hilberman, 'STAR: An active biquadratic filter section,' *IEEE Trans. Circuits and Systems,* vol. CAS-22, pp. 115-121, Feb. 1975.

[FT] P. E. Fleischer and J. Tow, 'Design formulas for biquad active filters using three operational amplifiers,' *Proc. IEEE,* vol. 61, pp. 662-663, May 1973.

[GG] F. E. J. Girling and E. F. Good, 'Active filters–12, The leap-frog or active ladder synthesis,' and 'Active filters–13, Application of the active ladder synthesis,' *Wireless World,* vol. 76, pp. 341-345, July 1970, and pp. 445-450, Sept. 1970.

[Go] J. Gorski-Popiel, 'RC-Active synthesis using positive-immitance converters,' *Electronics Letters,* vol. 3, pp. 381-382, Aug. 1967.

[Gr] A. J. Grossman, 'Synthesis of Tchebycheff parameter symmetrical filters,' *Proc. IRE,* vol. 45, pp. 454-473, April, 1957.

[Gu] E. A. Guillemin, *Synthesis of Passive Networks,* John Wiley and Sons, Inc., New York, 1957.

[HBG] B. J. Hosticka, R. W. Brodersen, and P. R. Gray, 'MOS sampled data recursive filters using switched capacitor integrators,' *IEEE Journ. Solid-State Circuits,* vol. SC-12, pp. 600-608, Dec. 1977.

[Hu] G. Hurtig, III, U. S. Patent 3,720,881, March 1973.

[Ka] T. Kailath, *Linear Systems,* Prentice Hall, Englewood Cliffs, NJ, 1980.

[KHN] W. J. Kerwin, L. P. Huelsman, and R. W. Newcomb, 'State-variable synthesis for insensitive integrated circuit transfer functions,' *IEEE Journ. Solid-State Circuits,* vol. SC-2, pp. 87-92, Sept. 1967.

[Ma] K. Martin, 'Improved circuit for the realization of switched-capacitor filters,' *IEEE Trans. Circuits and Systems,* vol. CAS-27, pp. 237-244, April 1980.

[Mi] S. K. Mitra, *Analysis and Synthesis of Linear Active Networks,* John Wiley and Sons, Inc., New York, 1969.

[MW] The MathWorks, Inc., *The Student Edition of MATLAB, Version 4 User's Guide,* Prentice Hall, Englewood Cliffs, NJ, 1995.

[Ri] R. H. S. Riordan, 'Simulated inductor using differential amplifiers,' *Electronics Letters,* vol. 3, pp. 50-51, Feb. 1967.

[Sa] W. Saraga, 'Sensitivity of 2nd-order Sallen-Key-type active RC filters,' *Electronics Letters,* vol. 3, pp. 442-444, October 10, 1967.

[SK] R. P. Sallen and E. L. Key, 'A practical method of designing RC-active filters,' MIT Lincoln Laboratory Technical Report No. 50, May 6, 1954; also *IEEE Trans. Circuit Theory,* vol. CT-2, pp. 74-85, March 1955.

[Sto] L. Storch, 'Synthesis of constant-time-delay ladder networks using Bessel polynomials,' *Proc. IRE,* vol. 42, pp. 1666-1675, 1954.

[Su1] K. L. Su, *Active Network Synthesis,* McGraw-Hill Book Company, New York, 1965.

[Su2] K. L. Su, *Time-Domain Synthesis of Linear Networks,* Prentice-Hall, Englewood Cliffs, NJ, 1971.

[Su3] K. L. Su, *Handbook of Tables for Elliptic-Function Filters,* Kluwer Academic Publishers, Boston/Dordrecht/London, 1990.

[Sz] G. Szentirmai, 'Synthesis of multiple-feedback active filters,' *Bell System Tech. Journ.,* vol. 52, pp. 527-555, Apr. 1973.

[Thn] W. E. Thomson, 'Delay networks having maximally flat frequency characteristics,' *Proc. IEE,* pt. 3, vol. 96, pp. 487-490, 1949.

[Ths] L. C. Thomas, 'The Biquad: Part I - Some practical design considerations,' *IEEE Trans. Circuit Theory,* vol. CT-18, pp. 350-357, May 1971.

[To] J. Tow, 'Active RC filters - A state-space realization,' *Proc. IEEE,* vol. 56, pp. 1137-1139, June 1968.

Index